T0140234

Natural Computing Series

Series Editors: G. Rozenberg
Th. Bäck   A.E. Eiben   J.N. Kok   H.P. Spaink

Leiden Center for Natural Computing

More information about this series at http://www.springer.com/series/4190

Mike Preuss

Multimodal Optimization by Means of Evolutionary Algorithms

 Springer

Mike Preuss
Lehrstuhl für Wirtschaftsinformatik und Statistik
   Westfälische Wilhelms-Universität Münster
Münster, Germany

Series Editors
G. Rozenberg (Managing Editor)

Th. Bäck, J.N. Kok, H.P. Spaink
Leiden Center for Natural Computing
Leiden University
Leiden, The Netherlands

A.E. Eiben
VU University Amsterdam
The Netherlands

ISSN 1619-7127
Natural Computing Series
ISBN 978-3-319-79156-2   ISBN 978-3-319-07407-8   (eBook)
DOI 10.1007/978-3-319-07407-8

Printed on acid-free paper

Springer International Publishing AG Switzerland is part of Springer Science+Business Media (www.springer.com)

to Carla

Foreword

Half a century ago, with more and more computers available at university and research labs, numerical optimization became en vogue. Direct search methods like those of Rosenbrock, Nelder & Mead, and M.J.D. Powell, to name just a few, helped to solve a lot of nonlinear, analytically intractable problems approximately. Unfortunately, such hill climbers found their way to only one local minimum or maximum in the vicinity of the given starting points in the search space. The case of multiple local optima seemed to be better treated by population-based methods like evolutionary algorithms, including genetic algorithms. Numerous experiments with properly tuned internal parameters of such black-box methods, as they were called, were published to demonstrate the suitability of those bio-inspired search procedures in multimodal landscapes. But there was no proof, no guarantee to find the optimum optimorum or global optimizer. Even worse, no satisfactory definition of all problems occurring in the case of multiple local optima existed. This situation remained the same for many years despite dozens of international conferences in the field of evolutionary and natural computation.

Mike Preuss' book is the first comprehensive treatment of many problems to handle multimodal optimization tasks by means of evolutionary algorithms in a structured manner. Indeed, there are a couple of different aspects to be obeyed when not only one local extremum exists. Sometimes only the global optimum is wanted, but it can be located at several locations (optimizers). Otherwise, all or only some of the optima are wanted. If not all of them, under which criteria should one select them? The author is probably the pioneer to create a taxonomy of the multitude of possible situations existing in multimodal optimization. One rather old idea in the field is niching: A population of seekers is split into subgroups each searching only in a subspace of the entire search space. There have been numerous such attempts, most of which are mentioned, characterized and evaluated, i.e., criticized. So far no satisfactory theory in the area of niching exists. This work is the first and so far only one to evaluate niching strategies rigorously to find out which ones are appropriate for which purpose. When one has found the promising basins of attraction of hopefully just one local extremum each, it is usual to start a traditional local optimum search, in this case one of the currently most successful evolution strategies (ES). Mike Preuss has combined his favorite basin detection method with such modern ES versions and sent them to benchmark-assisted international competitions—and won! Besides all previously mentioned leading-edge features of the book, this fact should attract interested readers particularly.

Hans-Paul Schwefel

Dortmund, Easter 2014

Foreword

Optimization problems arise in a wide variety of areas ranging from production, logistics, biology, and medicine, to engineering. The task in optimization is to find a solution, that is, an assignment of values to specific decision variables that gives the best possible value for a given objective function. In many cases, finding an optimal solution is a very difficult task due to nonlinearities in the objective functions and the possible occurrence of (many) locally optimal solutions that trap the search process. If finding a single optimal solution is already difficult, finding several or all optimal solutions is even more difficult; and it is this latter, so-called multi-modal optimization task that the author tackles in this book.

The focus in this book is apparently on evolutionary algorithms as solution methods and black-box continuous function optimization as the more specific problem class. This focus may, at first sight, limit the contributions of the work to these specific areas. However, this is not really true as the book contains a large number of more generic results and insights that make it relevant also beyond the field of evolutionary algorithms. In fact, any heuristic method for multimodal optimization should profit from techniques to identify basins of attraction of optima to make the search process more efficient, that is, from the techniques that are analyzed and designed here. One part of the contributions of this book develops formal models to analyze from a theoretical perspective the potential impact such techniques may have. This analysis is particularly interesting as it relates properties of the search space to the potential advantages of the considered techniques. While in the theoretical analysis specific techniques for basin identification and other tasks may be modeled, when it comes to actually solving multi-model optimization problems, effective algorithmic techniques need to be designed for implementing them. Another, generic contribution of the book is the development of the nearest-better clustering method for basin identification. This method is then used as a supporting tool for evolutionary algorithms for multi-modal optimization; however, it is directly applicable also to improve other heuristic search techniques. (Actually, the used evolutionary algorithms, in particular, the well-known co-variance matrix evolution strategy, could also be seen as efficient stochastic local search heuristics for black-box continuous optimization giving evidence for this claim.) The resulting multi-modal optimization method is particularly effective and shows excellent performance. This is confirmed by the fact that the resulting algorithm was the top-performer in a recent benchmark competition on multi-modal optimization.

Apart from these contributions, I would like to highlight two main additional ones. The first one is that the author has a consistent personal view of the research on multi-modal optimization and clearly organizes the contributions described so far in the literature. This may seem a minor contribution at first sight, but in the context of multi-modal optimization it becomes an important one as (i) many contributions have been obtained within different fields and many researchers are apparently not aware of the existing links, and (ii) many notions such as niching are used in very different senses and therefore lead to confusion even inside the same research community. The second contribution concerns the experimental evaluation. Unfortunately, in the history of evolutionary computation and, more generally, heuristic search algorithms, a sound experimental methodology has not always received the attention that is actually required. This book is exemplary in the adoption of a sound experimental methodology (which actually the author has helped to develop) and it will hopefully help to convince fellow researchers to adopt such methodologies in their own research.

In conclusion, I think that this book contains a large number of in-depth research results, and if multi-modal optimization is your research subject, this book is clearly a milestone that has to be read. In addition, the book provides a wealth of additional contributions that will make it an enjoyable and beneficial read even beyond the particular research subject treated. I therefore wish the book all the deserved success and a large future audience.

Thomas Stützle

Brussels, June 2014

Preface

This book is the result of a *very* long journey into optimization, and, more specifically, into *evolutionary computation*. This journey would not have been possible without the support of my family. I am very grateful to my parents Herbert and Christa, my sister Jennifer and her family, and of course to my daughters Janinka and Merle. Of course there are many more people who acted as signposts and/or motivators, and they shall be mentioned as well. In order to avoid a boring long list, I will try to wrap their names into a short chronological report before giving an overview over the book itself.

The first event that connected me to (evolutionary) optimization was a radio feature I heard soon after starting my studies. It dealt with optimization by adaptation of concepts from nature carried out by Hans-Paul Schwefel and the people at his Chair of Systems Analysis at the TU Dortmund. Joachim Sprave then led me into the world of parallel evolutionary computing, and provided me with a very important insight: not everything that is written in a book is right just because it is printed. In this environment, I first met Beate Bollig, who later on consistently reminded me to finish my dissertation, up to when this was actually the case.

After going international (EU project: DREAM) due to Thomas Bäck, I had the chance to experience my first real scientific cooperation, for which I have to thank Márk Jelasity, Gusz Eiben, Ben Paechter, and Marc Schoenauer. Meanwhile, criticism of the experimental evaluation of optimization algorithms was on the rise. Thomas Bartz-Beielstein and I teamed up for years in order to provide techniques and guidelines for countering this criticism. Chapter 2 is my view on experimental work and I applied the described methodology to all experiments in this book.

With Günter Rudolph and Boris Naujoks, I explored the foundations of multi-objective optimization and learned how to deal with engineers in several real-world optimization projects. Ruxandra and Catalin Stoean are my long-term Romanian connection, together we have challenged many interesting problems and developed nice algorithmic techniques for multimodal optimization and evolutionary support vector machines. Of all the people enlisted here, Ofer Shir is probably the one I have collaborated with who was or is most devoted to niching in evolutionary algorithms. We tried to define niching and multimodal optimization at a time when the second term was not yet commonly in use.

Heike Trautmann and I first met in Singapore, only to find out that we worked in related fields at the same university. Besides steadily pushing me to finish my dissertation, she enriched my life in many ways, not the least of which is a much stronger inclination towards statistical techniques. With Jens Jägersküpper, I undertook an exciting excursion into theory, or rather, algorithm engineering. Catherine McGeoch taught me to pose the right questions, and from Thomas Stützle I learned that evolutionary algorithms are not always the answer, but often a useful start.

This work would have been finished much earlier if I had not been sidetracked by the fascinating world of game AI. While starting research in this area together with Simon Wessing and Jan Quadflieg, I had the opportunity to meet Julian Togelius and Georgios Yannakakis, and, a little bit later on, Paolo Burelli. I owe a lot to all five of you! That it could be finished at all is probably due to Hans-Paul Schwefel, who provided encouragement when it was needed, and Simon Wessing, Bernd Bischl, and Günter Rudolph, who helped me to resolve the last important questions.

Last but not least, I would like to thank Ronan Nugent for recognizing the scientific contribution of this book, and for guiding me through the publishing process.

What is this book about? The field of multimodal optimization is just forming, but of course it has its roots in many older works, namely niching, parallel evolutionary algorithms, and global optimization. My aim is to bring all these together and thereby help to shape the field by collecting use cases, algorithms, and performance measures. In my view, it is very important to exactly define what the goals of such an optimization process are and also to obtain a good understanding of what the algorithms actually do during this process, especially with respect to the properties of the tackled optimization problems. More concretely, the main objectives of this work are listed in Sect. 1.4.

The algorithms I provide for basin identification and optimization are meant as a step forward, not as a definitive answer. I presume that there is still a lot of yet undiscovered potential in research on multimodal optimization, and I would like to encourage more research in this area.

Concerning the structure and usage of this book, the reader may find Sect. 1.5 useful, it contains a short description of the chapters and indicates which parts may be most interesting when addressing the different aspects of multimodal optimization.

Have a fun!

Mike Preuss

Bochum, July 2014

Contents

Nomenclature

γ	Euler-Mascheroni constant, approximately 0.577
λ	number of offspring individuals generated in one generation
μ	number of individuals who survive selection and are parents for the next generation
σ	step size/mutation strength
σ^0	initial step size/mutation strength
A	desired set of coupons
b	number of basins an (abstract) optimization problem possesses
c	number of basins covered by an algorithm at a certain time
D	number of search space dimensions of the treated problem
$d(\mathbf{x}_1, \mathbf{x}_2)$	distance between two search points
$f(\mathbf{x})$	objective (fitness) function (value) of search point $\mathbf{x}$
f^{*G}	global optimum, function value of (a) global optimizer
$G(r)$	distribution function for the nearest neighbor distance r
k	number of selected neighbors
P_t	population of search points at time t
$p_{BI}(\mathbf{x}_1, \mathbf{x}_2)$	probability of correctly identifying that two search points are located in the same basin
$p_{BR}(\mathbf{x}_1)$	probability of correctly detecting that the basin of a search point has already been found
R	redundancy factor
t_1	point in time during a heuristic optimization process when the first result can be delivered (usually very early)
t_2	first hitting time of the global optimum in a heuristic optimization process
t_3	latest first hitting time of all basins in a heuristic optimization process: all basins are discovered
t_c	cycle time, the number of repetitions before a random process arrives at the same state again
X	the whole search space
Z_η	expected waiting time for drawing n of a fixed set of coupons
$\mathbf{x}$	coordinate vector in the search space that determines a search point
$\mathbf{x}^{*L}$	local optimizer
$\mathbf{y}$	set of values for an abstract algorithm performance measure
$\mathscr{B}$	basin system, consisting of single basins B_i
$\mathscr{C}$	clustering, consisting of clusters (subsets C_i)

$\mathscr{D}$	set that contains all decided clusters of a clustering
AAHD	augmented averaged Hausdorff distance
AE	algorithm engineering
AHD	averaged Hausdorff distance
all-global	the target of the optimization is to detect all global optimizers
all-known	the optimization shall detect all existing optimizers, global and local
AOV	average objective value
APD	augmented peak distance
BA	basin accuracy
basin (of attraction)	the search space area from which a local search algorithm converges to a (local) optimizer
basin identification	detect the locations of the different basins of an optimization problem by identifying which search points belong to which basins
basin recognition	decide if the basin a search point belongs to is already known
BBOB	black box optimization benchmarking, an "*instance*" of COCO, held in the form of GECCO workshops (up to now 2009, 2010, 2012, 2013, 2015)[1]
BFGS	quasi-Newton method named after its inventors Broyden, Fletcher, Goldfarb, and Shannon
BFS	breadth-first search
BIPOP-CMA-ES	CMA-ES variant splitting the search effort between small and large, increasing population sizes
black box optimization	optimization without any knowledge about the system that generates objective function values, e.g., no analytical form or derivatives are given
BR	basin ratio
CCP	coupon collectors problem
CEC	annual (international) conference on evolutionary computation
ceteris paribus conditions	the experiment is repeated under exactly the same conditions, except for the starting time
CI	computational intelligence
CMA-ES	covariance matrix adaptation evolution strategy, introduced in Hansen and Ostermeier [103] and in details further developed since
COCO	comparing continuous optimizers, a platform for comparison of real-parameter global optimization algorithms, see http://coco.gforge.inria.fr/
COGA	cluster-oriented genetic algorithms
CSR	complete spatial randomness
DACE	design and analysis of computer experiments, deterministic precursor of SPO
dADE/nrand/*	DE/nrand/* with an additional dynamic archive
DBF	detected basin fraction
DE	differential evolution
DE/nrand/*	differential evolution variant that uses nearest neighbors as base vector for generating offspring
DECG/DELG/DELS	different differential evolution variants that emphasize parallel local searches
decided cluster	cluster of which the majority of constituents are located in the same basin (its main basin)
design	a set of design sites
design site	equivalent to experimental unit, here meaning the point in the algorithm parameter space that is tested

---

[1] web page of the 2015 issue: http://coco.gforge.inria.fr/doku.php?id=bbob-2015

DFS	depth-first search
DMM	detect-multimodal, short name for the hill-valley method
DOE	design of experiments, a set of techniques for setting up experiments, first introduced by Fisher [81]
DPI	dynamic peak identification
EC	evolutionary computation
ELA	exploratory landscape analysis
epistasis	related to separability, but defined over binary spaces: one phenotypical attribute is influenced by several genes or vice versa
ERT	expected running time
ES	evolution strategies
ETP	empirical tuning potential
F-Race	parameter tuning method
FMPM	funnel-based extension of the MPM generator
freestanding cluster	decided cluster whose main basin is different to all other clusters
GA	genetic algorithms
GECCO	annual (international) conference on evolutionary computation
GLOBAL	2-phase global optimization algorithm that employs single-linkage clustering in the global phase and BFGS for local seaches
global optimizer	location (possibly one of several) in the search space for which the objective function returns the global optimum
global optimum	best numerical value that is returned by an objective function
good-subset	the optimization shall detect a small subset of very good optimizers that are well spread over the search space
GP	genetic programming
hill-valley method	mechanism for detecting if two search points reside in the same basin by placing at least one point between them
ILS	iterated local search
IPOP-CMA-ES	CMA-ES variant with increasing population size
LHS or LHD	Latin hypercube sampling, Latin hypercube design, space-filling sampling method used within SPO as alternative to purely random (MC) sampling
local optimizer	location in the search space that corresponds to a local optimum
local optimum	objective function value of a point in search space that cannot be improved by making an infinitesimal step in any direction (note that this includes global optima)
locality principle	search points in the direct vicinity shall be more similar to each other than to more distant search points (in terms of objective values)
MC	Monte Carlo, meaning that a process (e.g., a sampling process) works completely at random
mixed cluster	cluster that does not have the majority of constituents in any single basin
MPM	multiple peaks model, test problem generator with randomly placed peaks
multimodal	objective function with at least 2 global optimizers
multimodal optimization	detecting several optimizers of a multimodal problem at once
multimodalCutProbs	NEA1/NEA2 parameter that determines how the DMM (hill-valley) method is used
NBC	nearest better clustering, a topological clustering method that makes use of objective values of a population next to the search space locations
NBC-CMA-ES	early version of the NEA1 algorithm

NEA1	niching evolutionary algorithm 1, first of two niching algorithms suggested by the author, basically a combination of initial random sample, NBC and CMA-ES, employing several populations concurrently
NEA2	niching evolutionary algorithm 2, suggested by the author, similar to NEA1 but doing local searches sequentially
niching (in optimization)	method to (implicitly or explicitly) recognize different basins of attraction and inject this information into an optimization algorithm
NND	nearest neighbor distance
one-global	the optimization shall find one global optimizer as fast as possible
optimizer	in the optimization context usually meant as *local optimizer*
optimum	contrary to common language, in the optimization context this is often understood as *local optimum*
PA	peak accuracy
ParamILS	iterated local search applied to the (algorithm) parameter space
PD	peak distance
PR	peak ratio
PSO	particle swarm optimization
QABR	quantity-adjusted basin ratio
QAPR	quantity-adjusted peak ratio
QMC	quasi-Monte Carlo, meaning that a deterministic process is used to emulate MC behavior
R5S	representative 5 selection
redundancy factor	ratio of actually performed local searches to necessary local searches (number of basins b)
REVAC	relevance estimation and value calibration, a tuning method
rule 2	extension of the NBC clustering method that takes the indegree of nodes in the nearest-better graph into account
SD	sum of distances
SDNN	sum of distances to nearest neighbor
search point	a location in the search space, in the real-valued case of zero volume, associated with at least one objective value
separability	separable functions can be solved by decomposing them into D 1-dimensional functions and aggregating the obtained optima, there is no interaction between the different variables
sigmaToDistance	NEA1/NEA2 parameter that controls how the step size is regulated according to the estimated basin size
SPD	Solow-Polasky diversity
SPO	sequential parameter optimization, model-based parameter tuning approach
sqr	semi-quartile range
surrounded cluster	decided cluster with the same main basin as at least one other cluster
TolFun	parameter of the CMA-ES stopping rules that refers to differences in objective function values
TSC/TSC2	topological species conservation algorithm
TSP	traveling salesperson problem, typical combinatorial optimization test problem
UCF	useful cluster fraction
unimodal	objective function with only one global optimizer
w.l.o.g.	without loss of generality
weak local optimum	the optimum does not correspond to a single search point but to a set of search points (e.g., a line or a plateau)

Chapter 1

Introduction: Towards Multimodal Optimization

Here we isolate the matter of this work within the large domain of optimization. We introduce a number of basic terms and algorithmic techniques in Sect. 1.1, prior to discussing different possible general aims of multimodal optimization in Sect. 1.2. Next, currently available evolutionary algorithms for multimodal optimization are discussed in Sect. 1.3 with the objective of establishing an improved taxonomy for these methods. Finally, Sect. 1.4 establishes the overall aims of the subsequent chapters.

1.1 Optimization and the Black Box

What properties do we attribute to black box problems and their solvers?

Optimization in general is the attempt to find a state of a system that is better in quality than states known beforehand. In the best case, this new state is optimal, so that it represents the best of all permitted states. To be optimizable, a system must have input variables whose values at least partially determine its output and which can be adjusted by the utilized optimization method. Under a single objective, quality is given by one output variable, which we assume here has to be minimized. Of course, one would be most pleased to find an input variable vector that leads to the optimal output value, but in practice (e.g., design optimization or computationally expensive simulation systems) one is often satisfied with any improvement over the currently known best solution. Additionally, when the best possible output value, the global optimum, is unknown, recognizing it—when it is found during optimization—is of the same complexity as solving the optimization problem itself. If nothing is known about the system to be optimized apart from the types and the number of input and output variables, the scenario is regarded as *black box optimization*. In such a situation, the input-output relation cannot be expressed in a closed algebraic form, and no derivatives of any order are available.

1

1.1.1 Objective Function and Global Optimum

Despite the presumed absence of a concrete function in the general case, it makes sense to treat the optimized system as if such a function exists. The function value would then be generated by simulation or even by running an experiment, which is termed experimental optimization. Thus, we introduce a formalism that may be utilized in a mathematical context: the objective function f. It is defined over the not necessarily finite set of permitted input variable vectors $\mathbf{x}$ (*search points*) of the search space X. For all practical purposes, i.e., real-world applications, we may assume that the set of function values $f(\mathbf{x})$ is bounded from below, and we are able to compute f only with finite (known) precision, so that the wanted global minimum, denoted by f^{*G}, exists (Equation 1.1). If solutions at infinity are excluded, the search space X is a closed, bounded set, and under the additional assumption that f is continuous over X, this is guaranteed by the extreme value theorem of Weierstrass.

$$f^{*G} = \min\{f(\mathbf{x})|\mathbf{x} \in X\} \tag{1.1}$$

In contrast to the uniqueness of the global minimum, there may exist multiple or even infinitely many search points corresponding to it. We call these *global minimizers*. However, without additional knowledge concerning the objective function representing the black box, it is impossible to determine the cardinality or the search space proportion of the set of global minimizers a priori. Nevertheless, we can expect a decisive impact on the hardness of an optimization problem. In the following, we will use the term *fitness value* or simply *fitness* synonymously with 'objective function value', as this nomenclature is common in *evolutionary computation* (EC). Furthermore, the process of attaining a fitness value is called *function evaluation* or just *evaluation*. The number of evaluations available until a solution for the optimization problem must be presented is often bounded, especially if these evaluations are costly in time, because they each require a call to a separate computer simulation.

For practical reasons, the notion of black-box complexity has been introduced for theoretical investigations of optimization algorithms: Except for evaluations, all operations are simply neglected during run-time analysis. Droste, Jansen, and Wegener [67] demonstrate how to utilize this concept for generalized randomized search heuristics in finite search spaces. Its roots go back to the investigation of oracles of zeroth order by Nemirovsky and Yudin [159]. However, note that this work deals with real-valued optimization problems in infinite search spaces. Nevertheless, dominating, constant costs of function evaluations are also often assumed in experimental research on optimization.

It may be noted that Equation 1.1 does not refer to constraints, which are usually formulated as separate functions $h_i(\mathbf{x})$. These are omitted here because constraints do not play a major role in this work.

1.1.2 The Locality Principle

At this point, it must be decided what further reasonable assumptions about the nature of f can be made. If we only accept knowledge about dimensionality and extension of the search space—which is often, but not always, constrained—and the existence of global minimizers, we are left with random search or total enumeration on a finite grid. More complex optimization methods are built on the expectation of finding more structure that may be exploited to speed up the search. The most important assumption needed for these algorithms is a certain smoothness, or, more generally, the validity of the locality principle: Fitness values of search points nearby are in general assumed to be more similar than the ones of search points located far away from each other. Here, the distance between two points has to be measured by means of a suitable distance metric.

When considering the immediate environment of a search point, the locality principle leads us to the notion of neighborhoods. However, the definition of a neighborhood strongly depends on the search space composition. In binary or discrete search spaces, it is common to define the neighborhood of a search point as the set of points reachable by taking the smallest possible steps in each direction. In real-valued search spaces, such a definition is questionable because it would relate neighborhood size directly to the representation of floating point numbers in a concrete machine. Although, in reality, steps would be larger than infinitesimal, they could still be much too small to be of interest for practical purposes. Instead, one may determine neighborhoods according to a distance metric by grouping all search points within a certain range together. Clearly, this method introduces an ambiguity because it utilizes a range parameter ε. Nevertheless, this construct is still useful because it embodies the prime reason for considering neighborhoods: the expectation that increasing the quality of a solution by taking small steps—and so using existing information—is more likely than by taking large ones. Equation 1.2 describes the ε-neighborhood of a search point $\mathbf{x}$ with regard to the distance metric d:

$$N_\varepsilon(\mathbf{x}) = \{\mathbf{x}' \in X : d(\mathbf{x}',\mathbf{x}) < \varepsilon\}. \tag{1.2}$$

1.1.3 Local Optimality

Until now, we have only considered the spatial distribution of search points, thus disregarding their associated fitness values. However, neighborhoods are also well suited to helping characterize another important feature of the objective function f: *local minimizers*. If for a neighborhood $N_\varepsilon(\mathbf{x})$ around $\mathbf{x}$, the best fitness value is $f(\mathbf{x})$, the search point $\mathbf{x}$ is a local minimizer, or, without reference to an optimization direction, a *local optimizer*. The associated fitness value $f(\mathbf{x})$ is then called *local minimum*, or, more general, *local optimum*. More formally, this is expressed in Equation 1.3 for real-valued search spaces. Note that increasing neighborhood sizes may hide previously detectable local minimizers.

$$\mathbf{x}^{*L} \text{ is local minimizer iff } \exists \varepsilon : \forall \mathbf{x} \in X : d(\mathbf{x}, \mathbf{x}^{*L}) < \varepsilon \Rightarrow f(\mathbf{x}^{*L}) \leq f(\mathbf{x}). \quad (1.3)$$

Admittedly, Equation 1.3 targets point-sized minimizers in the interior of the search space. Thus, there are two exceptional cases: line-shaped and areal local minima and minimizers on search space boundaries or discontinuities (ridges). The latter are covered by Equation 1.3, so that improper minimizers introduced by boundedness can be identified. However, the intuition behind the distance-based condition is that the search point in question is surrounded by an ε-environment in all directions, which is not possible if this would be cut by a search space bound. For the former case, plateaus of equal fitness, the equation leads to detection of an infinite number of minimizers. The associated local minimum is called a *weak local minimum*. Such a huge number of minimizers is clearly undesired, but consequences on the optimization process shall be limited. We may expect that plateaus on good fitness levels are reasonably small; otherwise the optimization task would be rather simple. Nevertheless, we shall keep in mind that Equation 1.3 does not allow us to determine whether two search points are located on the same plateau. For related definitions of local optimality and a discussion of optimum properties the reader is referred to the works of Boyd and Vandenberghe [39, p. 128], and Schwefel [205, p. 24].

The previously stated problem may seem artificial at first glance, but from a global perspective it has some relevance. We argue that the primary motivation for dealing with local optima comes from the observation that many optimization methods sooner or later converge to a single optimizer, be it a global or a local one. In particular, all *greedy* algorithms follow a descent, and are thus expected to do exactly this. Therefore, detecting search points on the same plateau or approaching the same optimizer surely helps speed up the search. Among others, Törn and Žilinskas [229] name this aim as the most important one when considering sophisticated global search algorithms. Every global optimizer is also a local one, but not vice versa. Thus, the only way to detect global optimality—with no additional problem knowledge given—is to visit many local optima, compare their fitness values, and select the best one as a potential global optimum.

Naturally, some objective functions do only possess one optimizer. These are generally considered easier to optimize and are called *unimodal*. All others are called *multimodal*—solving these is the focus of this work. Note that unless information concerning the number of optima is given, as is usually the case for benchmark objective functions, detecting more than just unimodality or multimodality may not be trivial.

1.1.4 Basins of Attraction

Inspecting many local optimizers in a fixed number of function evaluations requires effective means to determine whether different search points are located near the same local optimizer as early as possible. The local optimality condition of Equation 1.3 only regards the immediate environment of a local optimizer. But what about the interspace between these small areas? If we consider search paths of greedy optimization algorithms we may partition the whole search space into portions each containing the search points from which such an algorithm finally converges to a certain local optimizer. We term these sets *basins of attraction* or simply *basins*. Note that we explicitly tie basins to optimizers, not to optima, as opposed to the procedure advocated in [229]. This is motivated by the need to prevent non-compact basins resulting from distributed, but equally fitness-valued optimizers belonging to the same optimum. These would otherwise be combined into a single basin.

In real-valued search spaces, the basin of a local optimizer $\mathbf{x}^{*L}$ can be defined as the union of search points $\mathbf{x}$ for which a monotonic function α exists that describes a continuous search path from $\mathbf{x}$ to $\mathbf{x}^{*L}$.

$$\mathbf{x} \in \text{basin}(\mathbf{x}^{*L}) \Leftrightarrow \exists \alpha : [0,1] \to \mathbb{R}^n, \alpha \text{ continuous}, \alpha(0) = \mathbf{x}, \alpha(1) = \mathbf{x}^{*L}$$
$$\text{s. t. } f(\alpha(t')) \leq f(\alpha(t)) \, \forall t < t'. \quad (1.4)$$

This basin formalization is similar to the one employed by Addis [1] except for the allowance of monotonic instead of strictly monotonic paths. The change results in better compliance with Equation 1.3 by its merging the basins of the whole set of adjacent optimizers of a weak (areal) local optimum into one. Thus, any of these optimizers may serve as a substitute for all of them as far as their basins are concerned.

It should be noted that an unresolved issue remains when partitioning the search space using Equation 1.4: The obtained basins are not strictly disjoint — their borders overlap. The outermost points of each basin belong to at least one other basin as well. Törn and Žilinskas [229] therefore introduce the term *region* for sets containing these points, and delete them from the basins. In consequence, the union of all resulting basins is smaller than the search space itself. At least for relatively large search spaces, and assuming that we do not intend to use Equation 1.4 for mathematical proofs, we consider such a distinction unnecessary. It does not seem likely during an optimization with a fixed number of evaluations that many search points are positioned exactly on the boundaries of two or more basins. We may however run into problems if basin boundaries are not small but are areal structures (plateaus). Again, we assume that these cases are rare and should be dealt with on an ad hoc basis, e.g., by removing large plateaus from all basins and putting them into a newly created distinct one.

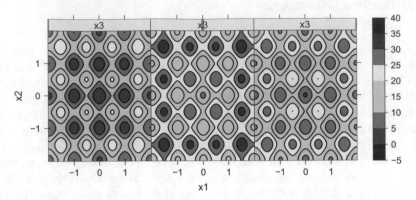

Fig. 1.1 Two-dimensional cuts through a three-dimensional Rastrigin function with $a = 5$. The third variable x_3 is set to (from left to right) 0, $\frac{\pi}{8}$, and $\frac{\pi}{4}$. Changing it obviously does not modify the basin structure but shifts the whole surface upwards and downwards.

For discrete search spaces, similar basin formalizations may be envisioned by giving up continuity of α and restricting its steps to minimal neighborhoods. In any case, the intuition behind it is kept unchanged: We partition the search space into areas in accordance with the behavior of a purely locally acting optimization algorithm.

1.1.5 Optimization Problem Properties

It may be asked how meaningful such partitioning is when considering optimization techniques which do not act strictly locally. Our answer is that all algorithms have to track down local optimizers, either implicitly or explicitly, to find the global optimum. While doing so, they may be susceptible to getting stuck in or wasting a large number of precious evaluations on a local one. This difficulty has often been referred to as the problem of *premature convergence*. Rudolph [192] has proved that even sophisticated algorithms such as *evolution strategies* (ES) with self-adaptation can be subject to it. From a different perspective and by using simple Markovian models, Schönemann, Emmerich, and Preuss [198] demonstrate that irreversible concentration onto one basin in evolutionary algorithms often happens much faster than would be expected. Thus, the number of basins surely is an important factor when considering the hardness of a problem.

Törn, Ali, and Viitanen [227] argue likewise, and additionally name two related properties, region of attraction size and embeddedness of the global optimum. The former corresponds to the basin size(s) of global optimizer(s) after Equation 1.4 . The latter refers to a structure on the basin level that permits educated guesses about the location of global optimizers. The well-known Rastrigin function $f(\mathbf{x}) =$

$na + \sum_{i=1}^{n} x_i^2 - a\cos(2\pi x_i)$, depicted in Figure 1.1, may serve as an example for a problem with an embedded global minimizer basin; in each dimension, it consists of a parabola with regularly distributed 'bumps' produced by the cosine term. For any optimization algorithm which is able to follow the parabolic structure despite the distortions introduced by local optimizers, solving the problem is relatively easy.

Another related regularity on the basin level that may be exploited to speed up optimization has been identified by several authors for specific objective functions: local optimizers that cluster into groups, so that it pays off to look for more optimizers in the vicinity of already found ones. Boese, Kahng, and Muddu [38] recognize a *big valley* structure in several combinatorial graph optimization problems, e.g., the traveling salesperson problem (TSP). Detected local optimizers form a valley (unimodal) structure in a small search space area; keeping initialization points for the proposed adaptive multi-start algorithm close to the best yet found optimizers proves advantageous. Reeves and Yamada [188] detect a similar big valley structure in flowshop scheduling problems and employ a specifically designed genetic algorithm (GA) to benefit from it. However, they find that this basin structure can be fragile and highly depends on the utilized search operators. Additionally, Whitley et al. [242] point out that small modifications to the optimized problem destroy the big valley.

For real-valued search spaces, similar compositions of basins into higher-level constructs have been observed in various real-world application domains, e.g., protein-protein interactions [233] and energy of atomic clusters [65]. Borrowing a term from biology, these clusterings of optimizers are named *funnel* structures; they can be exploited by smoothing the results of local searches and thereby approximating the assumed underlying function, e.g., by means of Gaussian kernel functions [3]. The funnel structures may also be understood as some kind of embeddedness in the previously described sense, but they additionally incorporate a strong assumption concerning the arrangement of optimizers. Considering only the discrete search space contingent made up of local optimizers, it is envisaged to possess a very simple, almost unimodal form. Both models require validity of the locality principle on the basin level.

Separability is a fourth important property. As Salomon [195] observes, many benchmark functions commonly used for scientific purposes are decomposable into D 1-dimensional functions, as is the Rastrigin function. Solving these separately is typically of much lower complexity than tackling the composed function. In contrast to this, objective functions of real-world problems are generally expected not at all to be decomposable. However, for an unknown objective function, we cannot presume full or no separability. Anything in between may be the case, with some interactions between variables being more important than others. Figures 1.1 and 1.2 demonstrate how such interactions can complicate the optimization of a previously rather simple function.

The related term for binary search spaces is *epistasis*. It originates from genomics, where it means that a phenotypical attribute is concurrently influenced by several genes and vice versa. In evolutionary computation, it usually refers to objective

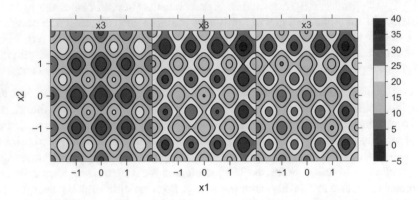

Fig. 1.2 Two-dimensional cuts through a three-dimensional modified Rastrigin function $f(\mathbf{x}) = 3a + \sum_{i=1}^{3}(x_i + x_3)^2 - a\cos(2\pi x_i)$ with $a = 5$. The change introduces interactions between variables x_1 and x_3, and x_2 and x_3, respectively. As in Figure 1.1, x_3 is set to (from left to right) 0, $\frac{\pi}{8}$, and $\frac{\pi}{4}$. Symmetry is lost for $x_3 \neq 0$, and the whole basin structure now moves.

functions for which the fitness contribution of individual binary variables depends on the values of other variables. See, for example, [59] for a treatment of epistasis built into benchmark objective function generators.

Summarizing, we have collected four properties beyond the obvious—the curse of dimensionality—that influence the hardness of optimizing a particular objective function. Many more, e.g. various forms of noise, may be important as well, but are not considered here to keep the focus tight. These properties are:

1. The number of basins; in the absence of plateaus, it equals the number of local optimizers.
2. Basin sizes of the sought-after optimizers relative to the extension of the search space; it is unclear how to handle this property in the case of unbounded spaces. But for the sake of simplicity, we may assume that such limits exist for the problems investigated here.
3. Existence of an exploitable basin-level structure, be it any form of embeddedness, big valley or funnel structure.
4. The degree of separability of the objective function.

Note that in 2, we replaced the global optimizers as used in [227] by sought-after optimizers. We stated that time may be a limiting factor and we have to accept suboptimal solutions as optimization results. In this case it is important to know if the treated problem exhibits a general trend towards smaller basin sizes for better optimizers.

1.1.6 Different Approaches

It appears pointless to attempt to enumerate even the most influential optimization algorithms ever investigated. Too vast is their number, too countless the research paths to be followed, and too scattered over various disciplines the work to be studied. Although the most important achievements of each 'school' may have crossed borders between disciplines, they seem to remain largely unconnected. Does this tell us anything besides the degree of diversification reached by modern science? A point we could make is that obviously still no universal optimization algorithm has emerged; nor does it appear likely that one ever will. Despite its weaknesses and limited applicability, the *No Free Lunch* theorem (NFL) [245] supports this view. We therefore restrict ourselves here to the approaches considered closely related to the evolutionary computation field, which is the main focus of this work.

When tackling a real-world optimization problem, one can generally follow one of two strategies or any compromise between them:

- either one takes the problem as given and adapts an algorithm to it,
- or one tries to find a "*good*" model/formulation of the original objective function that is easy to optimize by means of available algorithms.

The latter is often chosen in "*traditional*" mathematical optimization, as described by Boyd and Vandenberghe for convex optimization [39, p. 8]: "*If we can formulate a problem as a convex optimization problem, then we can solve it efficiently.*" But then one has to be aware that what is solved is the version of the problem given by the created formulation, not the original problem. It may require simplifications to obtain the formulation; these may cause unreachability of former local and global optima and so seriously alter the problem. At the same time, the formulation may introduce new optima, so that the solutions eventually found may be worthless in the context of the original problem. One may argue that any computational system that is derived from a real-world problem can only be a model of reality. However, the two approaches differ strongly in their priorities: make as few changes to the problem as possible vs. fit it to the relatively fixed demands of the employed algorithm.

For general nonlinear (more concretely, non-convex) optimization problems, the second ("*traditional*") approach is not applicable because no efficient general algorithms are known. Furthermore, in black-box optimization, no closed form of the problem is given, and simple mathematical objective function properties like differentiability or continuity cannot be taken for granted. Tightly related to the desire to optimize such "*unruly*" black-box problems is the term *direct search*, as it stands for abdication of mathematical analysis techniques within optimization methods. Algorithms of this type typically follow the first strategy and try to adapt to the optimized problem.

The term seems to originate in the 1961 paper of Hooke and Jeeves [107], but even in 2003, giving an exact definition of direct search methods was not trivial, as Kolda, Lewis, and Torczon [117] admit. Nevertheless, they also state that despite the difficulties of giving convergence proofs or other theoretical results for many of these methods, they have remained surprisingly popular with practitioners since their invention in the 1950s and 1960s. Although defining what direct search methods *do* may be difficult, characterizing what they *not do* seems to be agreeable: *"A direct search method does not 'in its heart' develop an approximate gradient"*, as Wright puts it [246]. This condition rules out quasi-Newton methods, but allows for other simple deterministic algorithms such as the *Nelder-Mead simplex algorithm* [158], which still belongs to the standard repertoire of optimization practitioners. Due to its heuristic nature, theoretical analysis proved difficult, so the few convergence results available, as presented by Lagarias et al. [122], are relatively new. This points to a general problem concerning most direct search algorithms: they often just work, but there is little theory. In other words, we do not currently understand very well why they work.

Besides seeing heuristic algorithms, the last few decades have also seen the invention of many stochastic methods, most of which mix determinism with controlled influence of randomness. While pure random search (the *Monte Carlo method*), the simplest of all optimization techniques, is universally applicable, it is also much too inefficient to be taken seriously. Probably, it was this picture people had in mind when they laughed at the new stochastic algorithms as reported by Beyer and Schwefel [31]. Nevertheless, the basic Monte Carlo procedure led to more successful methods, e.g., the *Metropolis algorithm* [149], *controlled random search* by Price [182], and *simulated annealing* by Kirkpatrick, Gelatt, and Vecchi [115]. In addition to physics, biology also provided inspiration for many direct search methods. *Evolutionary algorithms* (EAs) are a group of these, dating back at least to the 1960s. They will be reported on in Sect. 1.3. Others, more recently introduced, follow the *swarm intelligence* paradigm or model *artificial immune systems*.

Nowadays, it becomes increasingly difficult to rule out algorithms that employ randomness as not competitive just due to their probabilistic nature. A still growing subfield of computer science exclusively deals with *randomized algorithms* and utilizes them for optimization, cryptography, and other purposes. Motwani and Raghavan [155] provide a detailed overview. Interestingly, theory in this area is also growing, not least due to the large collaborative research project SFB 531[1] that took place at TU Dortmund from 1996 to 2008. This may be one of the reasons for the breakthrough of algorithms containing randomness in the 1990s.

---

[1] http://sfbci.uni-dortmund.de/

1.2 Multimodal Optimization

What exactly is the task when we speak of multimodal optimization?

When tackling a multimodal problem, we may address three related but different issues. These are:

1. Locate all optimizers corresponding to the global optimum.
2. Extract the full set of optima and optimizers the problem possesses.
3. Find the global optimum, together with at least one of its optimizers.

Clearly, these tasks are somewhat idealistic when it comes to real-world applications. If ever achieved, they are equivalent insofar as the *problem knowledge* needed to achieve one of them also enables us to achieve the other two easily. Once the complete set of optimizers has been found, attaining the global optimum and all its optimizers is a matter of a linear scan. Unless the set is complete, we cannot decide if it contains the global optimizer(s). But how realistic is such a scenario?

It goes without saying that detecting[2] a local optimizer—even if only approximately—needs at least one function evaluation. In a realistic setting, this 1:1 rate may well be reduced to 1:10 or even worse ratios. At the same time, the number of evaluations available for optimizing real-world problems is often bounded by relatively low values, today typically below 10^5. Many benchmark problems contain much higher counts of local optima, and it seems naïve to expect that practical applications are more simply structured.

Given that it might be impossible to find the global optimum, what remains of the previously stated tasks? Two different directions emerge. We may strive for locating as many local optimizers as possible, or concentrate search on the most promising regions first. These two approaches remind us of the two best known tree/graph search algorithms, *breadth-first search* (BFS) and *depth-first search* (DFS). With unlimited resources, the two ways reduce to one. Nevertheless, we argue in favor of the second approach since order of detection does matter when the optimization algorithm has to stop after processing only a small subset of the existing optima.

Li et al. [125] discuss the motivation behind multimodal optimization methods and name two desired effects: Firstly, the chances of locating the global optimum are increased. Secondly, attaining several good solutions, each with some distance in the search space, is appreciated in a design context (where good alternatives may have advantages not coded into the objective function and thus are invisible to the optimization algorithm). We assume that, for both effects, concentrating on the most promising regions first is an advisable approach. Additionally, Li et al. state that getting more than one good solution may improve our understanding of the objective function itself. When aiming at extracting knowledge, obtaining a broad spectrum

---

[2] Detecting here means testing (by computing function values) if it really exists. This excludes estimating optimizers based on a learned model we are confident in.

of optimizers may be helpful. However, in this work we give priority to searching for a small set of best possible solutions, if possible containing the global optimum. This is an obviously subjective choice that will be concretized further in Sect. 1.4. It is currently an open question how to trade off the quality and distance of the sought solution set. We will deal with approaches and performance measures for this problem in Chapter 5 and present some of our own results in Chapter 6.

The task of solving unconstrained, single-objective multimodal problems is closely related to two other branches of optimization. These are constrained optimization and multiobjective optimization, to both of which EAs are frequently applied.

In the former branch, it is quite common to get rid of the constraints by transforming the treated problem into an unconstrained problem by means of (metric) penalty functions (see, for example, Michalewicz and Schoenauer [150] and Coello Coello [50]). This method follows concepts previously outlined as *sequential unconstrained minimization techniques* (SUMT) by Fiacco and McCormick [80]. As multimodal objective functions are prevalent also in constrained optimization, the transformation results in an unconstrained multimodal problem.

In the latter branch, the focus has been mainly on the objective space for a long time. In that time, it has become increasingly clear that population movement in the decision (search) space heavily depends on the multimodal search properties of the applied optimization algorithms (Preuss, Naujoks, and Rudolph [174]).

1.3 Evolutionary Multimodal Optimization

What main EA variants for multimodal problems exist, and where do they originate from? What mechanisms do they employ to ensure global/parallelized search?

Evolutionary algorithms are well suited for solving multimodal problems as they are neither enumerative global nor strictly local search techniques. The interplay of local and global search in EAs is an important aspect when reviewing existing approaches, as done in this section. For multimodal problems, it is clear that global search abilities are of specific interest. Despite the fact that numerous EA variants have been proposed with the task of enhancing these, the big picture here is much more blurred than that concerning their local search capabilities. Additionally, theory is of little help because it is mainly devoted to the latter—this leaves us with models and experimental studies.

Apart from other optimization algorithms that originated in mathematics or *operations research* (OR), EAs are regarded as belonging to computer science (Figure 1.3)[3]. They are considered as an important building block of the *computational intelligence*

---

[3] This has been approved by the Informatik Duden [49] since its third edition (2001).

(CI) field established in 1994 at the first *World Congress on Computational Intelligence* (WCCI). Their focus is well summarized in the following definition:[4]

"These technologies of neural, fuzzy and evolutionary systems were brought together under the rubric of computational intelligence, a relatively new field offered to generally describe methods of computation that can be used to adapt solutions to new problems and do not rely on explicit human knowledge."

(David Fogel, [82])

Techniques that do not rely on explicit human knowledge have to be flexible. They must learn from online feedback and enable integrating experience gained during experimentation. The demand for flexibility may be one reason why so many different EAs exist today. However, many of the algorithmic details (e.g., concerning representations) of specific EA variants have their origin in the archetypal methods first invented.

1.3.1 Roots

Although there have been precursors to proposing the utilization of evolutionary concepts for optimization tasks, such as Bremermann [40] (also see Fogel's fossil record [83]), invention and development of the first evolutionary algorithms is nowadays attributed to a handful of pioneers who independently suggested three different approaches.

- Fogel, Owens, and Walsh introduced *evolutionary programming* (EP) [84], at first targeted at evolving finite automata, and later on modified into a numerical optimization method.
- *Genetic algorithms* (GAs), as laid out by Holland [106], first focused on combinatorial problems and consequentially started with binary strings, inspired by the genetic code found in natural life.
- Evolution strategies (ESs) as conceived by Rechenberg [187] and Schwefel [205] were designed for experimentally solving engineering problems such as designing nozzle contours [203] at first, and were applied to numerical optimization problems later on [204]. They are used with many representation types, the most prevalent of which is the vector of real numbers.

In the early 1990s, a fourth branch of evolutionary algorithms emerged, explicitly performing optimization of programs: genetic programming (GP), suggested by Koza [118]. Since about the same time, these four techniques have been collectively referred to as evolutionary algorithms, building the core of the *evolutionary computation* (EC) and *natural computing* field.

---

[4] Several such definitions exist; we chose one that is very close to our own opinion.

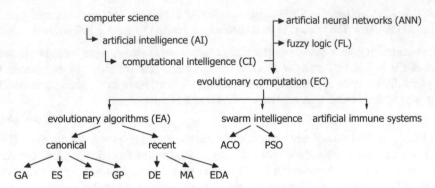

Fig. 1.3 Classification of computational intelligence, evolutionary computation, and neighboring paradigms within computer science. Only the most relevant technologies are depicted. The abbreviations employed in the leaves mean: GA = genetic algorithm, ES = evolution strategy, EP = evolutionary programming, GP = genetic programming, DE = differential evolution, MA = memetic algorithm, EDA = estimation of distribution algorithm.

It seems that the late 1960s and early 1970s was the right era for concurrent development of evolutionary optimization techniques at least three times. If we allow for interpretation of the progression of science as an evolutionary process, this coincidence may be described as a convergent evolution phenomenon. In biology, this term means the separate development of functionally similar body parts (such as eyes) or behaviors due to comparable selection forces. But what factors evoked the separate developments in this case? We could name at least three that may have contributed:

- Direct search methods, including those based on heuristics and randomness, were quite popular in the 1960s and 1970s (Sect. 1.1.6).
- The modern synthesis of genetics with (Darwinian) evolutionary theory had just become the standard model in evolutionary biology, and the 1950s and 1960s saw the beginnings of molecular biology, called the *molecular revolution* by Mayr [143].
- The first computers became available for scientific use, providing a huge speed-up in experimentation with simulation models and CAD.

As different as the original motivations for the four core algorithms had been, they still have much in common: Their creation was inspired by Darwinian thought and the rapidly progressing biology. Thus it is surprising that it took until the 1990s to bring the different communities together to realize that they were all heading in roughly the same direction and that the similarities found in the algorithms outweighed the differences by far. Now, we have a much more concrete idea of what these similarities are and hence what an evolutionary algorithm is.

1.3.2 The Common Framework

Fortunately, meanwhile, there is little doubt about the components and general structure of an EA. It is understood as a population-based stochastic direct search algorithm (not excluding population sizes of one, as featured in the simplest ES), that in some sense mimics natural evolution:

"An evolutionary algorithm (EA) is a metaheuristic optimization algorithm using evolutionary techniques inspired by mechanisms from biological evolution such as mutation, recombination, and natural selection to find an optimal configuration for a specific system within specific constraints."

(Wikipedia, [243])

The population aspect is missing in this otherwise concise definition; keeping sets of *individuals* (alternatively called *chromosomes*)—each containing information for one solution candidate—instead of one strengthens global search capabilities. In contrast to methods from mathematical optimization, population-based EAs may not start the optimization process with a single, given solution candidate; they can alternatively distribute the initial set of individuals over the available search space in various ways (see Morrison [154]).

Besides initialization and termination as necessary constituents of every algorithm, all 'pure'[5] EAs are composed of three important ingredients: a number of search operators, an imposed control flow (Figure 1.4), and a representation that maps adequate variables to implementable solution candidates.

Regardless of whether different EAs put different emphasis on the search operators mutation and recombination, their general effects are not in question. Mutation means neighborhood-based movement in the search space, which includes the exploration of the 'outer space' currently not covered by a population, whereas recombination rearranges existing information and so focuses on the 'inner space'. Selection is in most cases meant to introduce a bias towards better fitness values; GAs do so by regulating the crossover via mating selection, ESs utilize the environmental selection.

A concrete EA may contain specific mutation, recombination, or selection operators, or call them only with a certain probability, but the control flow is usually left unchanged. Each of the consecutive cycles is termed a *generation*. Concerning representations, it should be noted that most experimental studies are based on canonical forms such as binary strings or real-valued vectors, whereas many real-world applications require specialized, problem-dependent ones.

Although the described synchronous generation scheme is realized by the vast majority of EAs, there are some notable exceptions. The spatial predator-prey approach for multi-objective optimization as suggested by Laumanns, Rudolph and Schwefel [124] desynchronizes and localizes the selection process which is perfomed by

---

[5] neither hybridized with local search techniques nor extended by additional high-level structures such as multiple populations

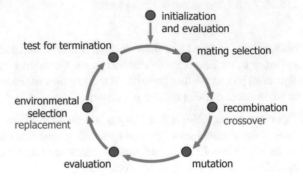

Fig. 1.4 The evolutionary cycle, the basic working scheme of all EAs. Terms common for describing evolution strategies are used, alternative terms are given below.

simulated predators that travel through a grid structure. Wakunda and Zell [239] and Runarsson and Xiao [193] propose continuous selection schemes, thereby removing the need to wait for termination of the evaluation of all offspring individuals. Such EA variants are especially well suited for an underlying parallel hardware as otherwise necessary synchronization points become obsolete.

For a detailed coverage on the defining components of EAs and their connection to organic evolution, see Eiben and Schoenauer [73] and Eiben and Smith [74].

1.3.3 Evolution Strategies

During the 1990s and in recent years, the three canonical EAs employed for numerical optimization, GA, ES, and EP,[6] experienced many attempts to import certain techniques from one main variant into another. At the same time, it was found that the three variants have far more similarities than expected, so exchanging operators is straightforward. The term 'EA' now basically means *any of GA, ES, EP, or combination thereof, possibly enhanced with other optimization techniques or problem dependent methods*. The traditional representation mapping, e.g., GAs for binary and ESs for real-coded problems, if ever valid (see, e.g., Schwefel [202]), has also been blurred and diminished to a rough guideline. Moreover, at least three new players have entered the scene in recent years, as shown in Figure 1.3, namely *differential evolution* (DE), *memetic algorithms* (MAs), and *estimation of distribution algorithms* (EDAs). It may be argued that DE variants basically resemble evolution strategies with a new variation scheme, and EDAs are in general very similar to other EAs with adaptation schemes. Memetic algorithms are EAs enriched by local search methods. However, even if the basic evolutionary principles remain unchanged, these methods are nowadays established and have their own communities.

---

[6] We do not consider GP here because it is not used in this context.

Does it make sense to nevertheless speak of an evolution strategy when determining the 'type' of an EA? What are the unique features that legitimize the term ES? Or, conversely, what are the implications of classifying any EA as an evolution strategy? We argue that the three following properties make the difference:

- Selection is deterministic (truncation), and the environmental selection step is strongly emphasized, compared to the mating selection phase.

- Self-adaptation or other parameter control mechanisms are prevalent for mutation strength variation during an optimization run. The appropriate control parameters are encoded into the chromosome/individual representation.

- With the exception of (1+1)- and (μ+1)-ESs, the steady state versions, the number of concurrently produced offspring individuals is significantly larger than the current population count (birth surplus), resulting in an easily adjustable selection pressure.

We abstain from a detailed discussion of different ES variants and operators apart from these properties and refer you to Beyer and Schwefel [31] instead. See also Hoffmeister and Bäck [105] for a comparison of basic mechanisms in genetic algorithms and evolution strategies. De Jong [57] provides a more general perspective. Note that, differently from evolution strategies, canonical GA variants with binary representation need fixed search space bounds and minimal step accuracies, whereas an ES can be employed as an unconstrained optimization algorithm.

As of 2014, it seems that the *covariance matrix adaptation evolution strategy* (CMA-ES), originally conceived by Hansen and Ostermeier in 2001 [103], is the predominant evolution strategy and probably the most used evolutionary optimization algorithm in the continuous domain. It has been developed further step by step to overcome apparent weaknesses (see the discussion in the introduction of [176]). The variant with increasing population size, IPOP-CMA-ES [14], is frequently used as the standard method to compare new algorithms.

1.3.4 EA Techniques for Multimodal Problems

Evolutionary algorithms have not been envisioned as local search methods and thus always possess global search capabilities to a certain extent. However, since about the 1980s, many EA variants have been suggested, especially for coping with highly multimodal problems. Interest in these may have been triggered by the rise of parallel hardware (see Alba and Tomassini [7]) on one hand and the intensifiying speciation debate in biology (Sect. 3.1) on the other hand.

It is not by accident that parallelism led to new, successful EAs for multimodal optimization; maintaining several distinct search paths—subsequently or concurrently—is a prerequisite for detecting multiple optimizers. If the pursued search paths correspond to different basins of attraction, every optimization run yields in finding several optimizers, local or global. Such information may be valuable by itself if multiple solutions are desired; additionaly, it enhances the chances of determining the global optimum of a problem (see Sect. 1.2). Note that parallelism as utilized by most EAs does not necessarily mean implementation on parallel hardware. Population-based optimization algorithms always inherently use parallelism even though they usually run on sequential computers. In fact, in the following, we make the assumption of a finite constant effort (e.g., number of evaluations) available to the whole optimization process because it matches best with the situation often found in real-world applications, where time often is the limiting factor and parallelization is difficult due to (simulation) software limitations.

Apart from its historical significance for EA development, the parallel stance may also provide keys for detecting differences and similarities between the existing algorithm variants. But in what ways do current EAs employ parallelism to strengthen global search capabilities? Surprisingly, there are only a few taxonomic attempts investigating this, namely the ones suggested by Ursem [237] and Eiben and Smith [74]. These utilize different, mutually incomparable criteria, such as the general strategy (avoid or repair) of algorithms in the former and *diversity maintenance* (implicit or explicit) in the latter. Both these categorizations sort the algorithms into tree-like structures, thereby implying that an unambiguous importance order of the differentiating criteria can be ascertained.

Here, we try an alternative approach by first establishing a set of unweighted properties describing actual algorithm behaviors rather than underlying ideas, shifting the focus from the functional 'how?' to the more abstract, purely observing 'what?' perspective. Note that properties do not simply correspond to classes of a classification in our view; any EA variant may possess any combination of properties. For a guideline to devising these properties, we look for different forms of parallelism (in the context of a sequential machine model), of which we find the following:

Parallelization in Time: This holds for all multistart methods, e.g., multistart hill-climbers and sequential niching, as proposed by Beasley, Bull, and Martin [26]. Potential solutions are obtained consecutively; every new instantiation may be provided with search results of completed previous runs. The available information is thus reliable but sparse at the beginning.

Parallelization in Space: All population-based methods share this property to a certain extent, but we especially focus on algorithms employing means to further structure the set of concurrently existing solution candidates, the most thoroughgoing of which are the parallel (independent) hillclimbers. Contrary to parallelization in time, there is much unreliable information concerning other search paths available from the start, getting more and more dependable as time progresses. Among others, Schütz and Sprave [200] early on employed a parallelization in space model with multiple populations placed on a toroidal neighborhood structure but with no strong interaction concerning the single search space areas.

Parallelization in Search Space: Without any knowledge about the structure of the search space, comparing relative or absolute distances of solution candidates and applying clustering methods are common means of preventing overlapping search paths and promoting good search space coverage. Techniques referring to the search space as relatively uniform are betokened as diversity maintenance methods. Diversity, understood as spread over large portions of the available search space, may be held up explicitly or implicitly. Following Eiben and Smith [74], explicit means that active measures are taken to model the distribution of search points in the desired way, whereas implicit means the deliberate slowing down of information exchange by restricting recombination or selection/replacement.

Parallelization in Basins of Attraction: Like diversity maintenance, this form of parallelism strives for a suitable spread of search points, only on a different level. The search space is not regarded as uniform, but as consisting of multiple basins of attraction. As Mahfoud [137] points out, it is the aim of *niching* algorithms to detect separate basins and keep them in the focus of the search. Unfortunately, *basin identification* within an EA is not easy and is prone to error (see Sect. 3), so that endogenously retrieved basin information—in contrast to problem knowledge, which may be used for evaluating algorithms—is highly unreliable and virtually nonexistent when the optimization starts.

Parallelization in Algorithmic Techniques: EA variants also differ in their utilization of auxiliary optimization techniques, namely local search methods. If any search point has arrived in a basin that is expected to feature a good optimum, global search capabilities may be given up in favor of a faster local search. Alternatively, hybridized algorithms that switch to the currently most appropriate technique during an optimization run (e.g., Krink and Løvbjerg [120]) may prove worthwhile. However, as with basin identification, successfully applying these techniques certainly requires well-developed diagnostic/analytic helper methods or a sophisticated learning scheme.

Fig. 1.5 Visualization of the proposed taxonomy, based on four different types of parallelization. Hillclimbers are marked in blue, EAs inspired by parallel implementations in red, panmictic EAs in orange, and niching methods in green. Differentiation expresses conceptual discrepancies rather than concrete values. The six algorithm types in the upper left corner are similar in their use of parallelization and thus represent one single point. Further distinction requires utilizing more than four properties or even specifying actual operators and parameter values.

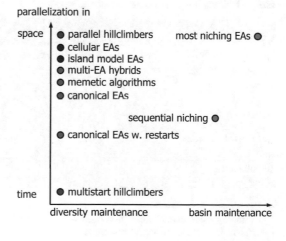

All but the last property may be quantified and thus utilized for visualization, as depicted in Figure 1.5. Furthermore, under the finite constant effort assumption, parallelization in space and time are clearly antithetic because they use up the same resources. Introducing restarts moves effort from space to time parallelism. Similarly, parallelization in search space and in basins of attraction is hardly achievable at the same time, as evenly covering the search space requires search point distributions different from the ones needed for evenly covering the basins of attraction.

The proposed scheme is far from complete. Many algorithmic features, such as unorthodox operators, e.g., the merging operator of *genetic chromodynamics* (GCD, [69]), remain untreated. However, it may suffice to attain an impression of the main streams of EAs employed for solving multimodal problems.

Table 1.1 comprises a lineup of the algorithm types from Figure 1.5 with Boolean values for all five parallelism-based properties. Although four of these are quantifiable in principle, assigning concrete absolute values appears inadequate without knowing concrete algorithms and parameters. Nevertheless, one can determine differences. For example, parallel hillclimbers and canonical EAs both provide parallelism in the search space. However, the former is expected to emphasize this property much more strongly due to the employment of non-unary recombination and selection operators in the latter.

The fifth property (parallelism in algorithmic techniques) enables us to partition the group of six EA types in the upper left region of Figure 1.5 into two fractions: multi-EA hybrids and memetic algorithms vs. canonical EAs, island model EAs, cellular/diffusion model EAs, and parallel hillclimbers. A criterion for further subdivision could be obtained from differences in the amount and structure of communication[7] between individuals. Model-based investigations concerning information speed have been undertaken, e.g., in Preuss and Lasarczyk [173], but

Table 1.1 How different algorithm types employ parallelism. Only conceptual differences are assessed without respect to concrete values.

algorithm class	parallelization in:				
	time	space	search space	basins	techniques
multistart hillclimbers	✓		✓		
parallel hillclimbers		✓	✓		
canonical EAs		✓	✓		
with restarts	✓	✓	✓		
memetic algorithms		✓	✓		✓
island model EAs		✓	✓		
multi-EA hybrids		✓	✓		✓
cellular EAs		✓	✓		
niching EAs		✓		✓	
sequential niching	✓	✓		✓	

---

[7] We employ the term *communication* for every type of information exchange between at least two individuals that results in altered genotypic composition of a population, e.g. due to recombination.

have provided only limited insight so far. A more sophisticated, unified approach by Sprave [216] (comprising canonical, island model and cellular/diffusion model EAs) solely focuses on selection, thereby neglecting partial information exchange due to recombination. Apparently, building a simple but meaningful model of the communication behavior of EAs is non-trivial.

Note that currently no general guidelines exist to govern selection of one of the algorithm types for application to a specific problem. Available comparison studies rarely comprise more than two or three of them and seldom refer to real-world test cases. Moreover, it is not even clear whether all these different algorithm types each correspond to problem classes in which they dominate all others. This is probably due to the fact that the algorithms were partly developed in parallel (rendering convergent evolution possible), mostly without the defining of a clear task other than that of optimizing multimodal problems. A step in this direction has recently been taken with the introduction of a data mining-based algorithm selection method termed *exploratory landscape analysis* (ELA), in Mersmann et al. [147], but the currently available results are very focused on a small set of benchmark problems.

1.4 Objectives of This Work

The aim of many past and current publications in the EC field appears to be to suggest new optimization algorithms that outperform the established ones on most or all available test problems. This may be especially tempting, considering that implementing a new idea or modifying an existing technique is straightforward in most cases and comparisons on benchmark test suites are easily done. But as previously stated, the NFL leads us to the conclusion that setting a too general task (optimizing all problems well) will most likely end up in failure. Taking into account that most heuristic optimization algorithms do not strictly adhere to the underlying assumptions of the NFL and thus could be more efficient if superfluous search steps were removed, it follows that a less ambitious task may be worthwhile considering. Instead of looking for the all-dominating optimization algorithm, we shift our focus to investigating what problem types any particular EA is good for. We suspect that pursuing this aim has two major advantages. On the one hand, well-performing algorithms for specific problem types may be obtained. On the other hand, separating the algorithms based on performance on different problems and detecting the underlying mechanisms affecting this performance may result in a better understanding of the algorithms themselves.

As this work explicitly deals with evolutionary algorithms for multimodal optimization, we consider the aforementioned (Sect. 1.3.4) parallelization properties as of particular importance. We will especially focus on the algorithms providing parallelism in basins of attraction, known as niching EAs, and their comparison to canonical EAs. The reason we prefer these to diversity maintenance type EAs is simple. Niching takes the characteristics of a treated problem into account, diversity maintenance ignores them and reacts only to modifications of the currently covered search space size.

More precisely, our aims are the following:

1. To improve the understanding of what niching in evolutionary algorithms actually is and what its potential benefits are.
2. To investigate whether niching techniques are suitable as diagnostic tools for experimental analysis, especially for detecting problem (type) properties.
3. To measure and compare the performances of niching and canonical EAs achieved on different benchmark test problem sets.
4. To make a move towards estimating for what problem types niching EAs actually outperform canonical EAs.

We will review our progress with respect to these directions in the final chapter, but first provide an overview of the structure of the whole book. We do so in particular to provide entry points for those who do not want to read it in full.

1.5 Book Structure and Usage Guide

This chapter has introduced the basics of evolutionary black-box optimization; it is the aim of the next one to provide foundations for our experimental work. As of now, there is little theory in multimodal optimization, so we need to base much of our reasoning on experimentation. After holding conference tutorials and other talks on how to perform experiments in the context of evolutionary computation for many years (mostly together with Thomas Bartz-Beielstein, starting with the CEC 2004 conference), the situation has certainly evolved. In most recent publications, we see the setup properly separated from the display of the results, and journals as well as conference review systems highly encourage us to provide enough detail to allow for replication of the results. So far so good?

The Bartz-Beielstein book of 2006 [16] provides an overview of the *sequential parameter optimization* (SPO) method, covering experimental setup, measuring, parameter tuning, and statistical evaluation. Chapter 2 is not intended as a revision course on this material. Instead, we want to emphasize the replicability aspect, which is, in our opinion, strongly connected to structured experimentation *and* its documentation. Consequently, we have applied the reporting rules (the whole methodology) to all experiments of this work. We would also like to recommend this to the reader: It can speed up writing down experimental results because it enforces structured thinking. And it also guides the reader, because it helps to prevent long, unstructured reports that can make detecting a setup detail very time-consuming. Of course, structured experimental reports are not a new invention. Natural sciences have used them for decades, if not centuries. For good reason, we presume.

Additionally, Chapter 2 discusses the similarities between parameter control and parameter tuning, and the adaptability aspect of parameter tuning. These cannot be found in [16] because they were published much later.

Chapter 3 turns to the existing niching definitions in evolutionary computation and their relation to the biological archetypes. We then set up a very simple model algorithm in order to explore the limit behaviors. This takes up ideas of [168] and extends them. Interestingly, some of the more simple cases can be computed instead of simulated due to their similarity with the *coupon collector's problem* (CCP), for which a number of theoretical and numerical works exist. We obtain some surprising results which are, after some consideration, well in line with the current research state in (evolutionary) global optimization and the younger field of multimodal optimization. For the former (one solution sought), niching, even if an ideal technique is available, it is not always helpful, at least not in the case of approximately equal basin sizes. For the latter (several solutions sought), it usually is. However, in the absence of a taxonomy of real-world problems, even if these are restricted to continuous optimization, we currently have little knowledge about how realistic the condition of nearly equal basin sizes is. Our guess is that it is rarely met.

In conclusion, this part lays the foundation for the methods and algorithms introduced in the later chapters; it is in some sense a justification for attempting niching and/or multimodal optimization. However, it is not necessary to read the chapter completely in order to understand the following ones.

Chapter 4 investigates a method and its extensions for what we call *basin identification*, the attempt to find out where in the search space the basins are actually located. If this can be achieved with sufficient accuracy, the results of Chapter 3 may be utilized to predict how sucessful a certain type of optimization algorithm will be. The basin identifiction runs on an initial sample that is usually available at the start of an evolutionary algorithm and thus before actually starting the optimization process.

The *nearest-better clustering* (NBC) method was introduced in [177] and extended in [171]. Here, we additionally suggest and evaluate a correction method for large samples (where the number of basins is usually overestimated) and then perform a large-scale parameter investigation of the NBC in order to derive good default values for the various settings (as the NBC is a geometrical method, it changes its behavior in different problem dimensions due to the *curse of dimensions*). The assessment is performed on a simple polynomial-based problem generator (MPM) that was originally suggested in [173] and is extended here in order to also represent funnel structures.

Note that NBC is relatively simple and can be embedded also in other population-based optimization algorithms without additional evaluations. The outcome (clustering with respect to actually existing basins of attraction) may help in setting parameters or even choosing algorithms that match the given problem well. Furthermore, NBC is in principle not dependent on a real-valued problem representation; it only needs a matrix of distance values between the evaluated search points. It may therefore also be applied to other problem types.

Chapter 5 first discusses the relationship between niching and multimodal optimization and then defines four use cases for the optimization of multimodal problems, depending on the solution type that is desired (one or many, global and/or local optima). It then discusses the available performance measures for multimodal optimization as enumerated in [180]. It turns out that the field is much more immature than one may think, because there are many reasonable measures of which only a few are currently widely used. There is undoubtedly a need for further exploration and evaluation here, but this is beyond the focus of this work.

The last part of this chapter reviews a large number of available methods for multimodal optimization, complemented by many that were originally designed in fields other than evolutionary computation. Especially the cluster analysis algorithms employed in global optimization in the 1970s and 1980s appear to be related to the NBC method we utilize. We suggest a separation of these methods into three classes, from explicit basin identification to simple diversity maintenance. This is a first attempt towards a taxonomy, which may, as with the described methods, be a source of inspiration for researchers in the multimodal optimization field.

Finally, Chapter 6 presents and experimentally evaluates our own two optimization methods (NEA1 and NEA2), both based on the NBC method for basin identification. A parameter investigation on problems from the BBOB collection and our own MPM/FMPM generator provides reasonable default values and also rules out some attempted extensions such as the use of the *detect-multimodal* (DMM) method. Finally, both algorithms are evaluated on multimodal problems from the BBOB set for the *one-global* case and on the problem set of the 2013 IEEE CEC niching methods competition for the *all-global* case. This competition was won by NEA2, and our analysis shows why it is better than the other algorithms: when only a moderate number of optima are sought, NEA2 is able to find most of these with very good accuracy due to the NBC clustering and the local search abilities of the CMA-ES. The detailed experimental description may be useful for other researchers who attempt to incorporate parts or techniques of our algorithms into their own algorithms.

The last chapter summarizes our findings and assesses if the goals stated in the introductory chapter have been fulfilled. It also provides pointers to the parts of this work that deal with each of the specific goals.

Finally, we make a statement concerning the use of the methods and findings of this work for solving real-world problems. In our experience, real-world problems are generally very different from each other, and a good solution to any single one relies on sufficient knowledge of the problem itself. Experiments with MMO methods may help in acquiring this knowledge, but this process is neither quick nor simple; one has to be ready to make lots of changes to the algorithms in order to be successful. This is clearly outside the focus of this work.

Furthermore, there is insufficient knowledge about the structure of and relations between real-world problems; there is no taxonomy that could guide algorithm usage. Most likely, it is possible to find a number of problems that can be solved well by the algorithms we investigate in this work (this approach would also work for most other optimization algorithms), but the results would only be transferable to very similar problems. We therefore call for more research on the interplay between different real-world problems, and problems and algorithms, but confine ourselves to establishing useful algorithms here.

Chapter 2
Experimentation in Evolutionary Computation

In which we reflect on the current status of experimentation in evolutionary computation (Sect. 2.1) and beyond (Sect. 2.2). We then argue in favor of a methodology in Sect. 2.3, highlighting the need for a structured approach with well-defined aims, parameter settings, designs, and measures. Finally, Sect. 2.4 deals with the positive aspect of parameters: the possibility of adapting algorithms to concrete needs via new, more suitable parameter settings.

2.1 Preamble: Justification for a Methodology

Why care about methodological questions concerning experimentation, especially in Evolutionary Computation?

At first glance, devoting a full chapter of this work to the experimental methodology seems to exaggerate its importance. Many older and some new scientific writings in the field of evolutionary computation have preferred an on-the-fly description of the employed experimental techniques or completely omitted such information. We argue that especially stochastic algorithms with many parameters require carefully designed and evaluated experiments. Although past studies in EC have often suggested new algorithms and experimentally substantiated their claim of suitability for a certain purpose, most of these have rapidly disappeared. It seems that the given evidence in these cases did not suffice in convincing other researchers to use and further develop the suggested algorithms. Why is this?

Experimental investigations of stochastic optimization algorithms face two major difficulties: nondeterminism and arbitrariness of environmental conditions. Nondeterminism is of course built into stochastic algorithms. It leads to varying results in quality and required effort and thereby severely complicates measuring and evaluating algorithm performance. More precisely, difficulties induced by observing the temporal dimension are twofold. Firstly, under ceteris paribus conditions (all else is equal, except for the starting time), even two runs of the same algorithm may show very different runtime behavior (see Hoos and Stützle [109]). Secondly, determining stagnation, or rather quasi-stagnation, of a stochastic optimization algorithm—the point in time when it is not going to achieve any more progress with a probability near 1—is usually nontrivial.

Arbitrariness, in turn, stems from the many degrees of freedom any experimental investigation of EAs possesses: There are several parameters to adjust, many test problems to choose from, and many reasonable performance criteria available. The only source of influence that may be safely disregarded is the effect of the employed hardware, as one usually confines oneself to targeting function evaluations as atomic units of measurement. Nevertheless, a realistic experimental setting may incorporate only a tiny subset of the possible setups, which is why the experimenter is forced to make many restrictive decisions before the first results are obtained. Although one often strives for proper justification of these decisions, some of them may have to be taken based on conjectures or simply personal preferences. We argue that it is often the expectation—or experience—of the reader, that out of the other setups that correspond to the experiment's basic assumptions, many would have led to qualitatively different outcomes.

Of the three named factors that may induce ambiguities, algorithm parameters, test problems, and performance criteria, the first one is of specific interest to algorithm engineers. In contrast to the other two, the parameters still stay in full control of the EA user even in the case of a real-world deployment, whereas problems and performance criteria may be regarded as given and relatively static. As the parameters can be interpreted as input variables, which surely have an impact on optimization algorithm performance, classical methods from statistics may be applied, namely *design of experiments* (DOE) techniques as first introduced by Fisher [81]. Standard textbooks like Montgomery [152] give a thorough introduction to these techniques; they thus help to establish a basic experimental methodology.

However, the last decade has seen a movement towards applying parameter tuning methods to evolutionary algorithms, intermingled with a greatly improved experimental methodology. We have been working together with one of the exponents of this movement, Thomas Bartz-Beielstein [16], and this chapter can also be seen as a personal summary of experiences with and insights into experimenting with (optimization) algorithms in the previous few years. In the following, we emphasize the general outline of the experimental methodology and some specific interest areas (e.g., adaptability) that may be further developed in the years to come. We deliberately leave out many details of the tuning process and the underlying modeling algorithms as these are of minor importance for the later chapters. However, the interested reader may find such information in the recent collection Bartz-Beielstein et al. [18], especially in the chapters [20] and [23], and in Bartz-Beielstein and Preuss [24].

2.2 The Rise of New Experimentalism in Computer Science

How does experimentalism evolve in other fields?
What is the situation in evolutionary computation?

In the last few years, theory on evolutionary algorithms has made astonishing progress. But still, most published work relies solely on experimentation as proof of concept or for performance comparison of different EA variants. However, despite around 40 years of experimental practice, there is yet no unified methodology for obtaining or even reporting experimental data. Nevertheless, we notice a certain convergence towards utilization of statistical techniques such as hypothesis testing. These are well in order to help judge the (statistical) significance of results, but cannot substitute asking scientific questions and reaching conclusions about what Mayo called *"severity"* (Mayo [140]) of outcomes: A demonstrated difference must also be scientifically meaningful to be of any value.

2.2.1 New Experimentalism and the Top Quark

Mayo herself is only one of the leaders of a recent multidisciplinary movement termed *new experimentalism*. Why should there be a need for a revival of experimentalism? Natural sciences have been the subject of theories, experiments, and basic forms of statistics for hundreds, if not thousands, of years. As an example, particle physics is a discipline with well-established, and sometimes even cyclic, interplay between theoretical and experimental practice: Theoreticians generalize gathered observations in mathematical models, and, based on these, make predictions, which in turn can be validated experimentally.

This interplay is impressively demonstrated by the case of the discovery of the top quark: the existence of the top quark had been postulated, based on the (theoretical) standard model, since the bottom quark was found experimentally in 1977. In 1994, 't Hooft and Veltman estimated its mass quite accurately at 145–185 GeV (which earned them the Nobel Prize in 1999). One year later, the top quark was discovered experimentally at Fermilab. Its real mass, drawn from current measurements of different teams, is now given as 172.9 ± 2.9 GeV [45].

Unfortunately, in life sciences as well as in the areas of computer science that have seen the application of heuristics or stochastic algorithms, a number of factors can complicate learning from experiment/observation. One counterexample does not falsify a conjecture, as the outcomes are not deterministic but rather samples from unknown distributions. Especially, performance assessment of stochastic optimization methods depends on many decisions (as argued in Sect. 2.1), some without quantifiable criteria available. Experimentation must *"define itself"* to be of a certain value. It is our opinion that two of the intentions behind new experimentalism are especially important to experiments in computer science:

1. Statistical techniques must not be applied *"blindly"* (statistical significance versus scientific severity).
2. Experimentation shall be the glue between theory and practice, preventing divergence of *"straightforward theory"* on the one hand, and *"real-life practice"* on the other hand.

2.2.2 Assessing Algorithms

Whereas the new experimentalists support empiricism per se and provide statistical tools to learn from experimental outcomes, their main ambition is not to outline concrete experimental procedures, as these would heavily depend on the investigated object and the concerned scientific field. We now move to computer science and review some contributions towards structured experimentalism on algorithms. Interestingly, the terms *"experiment"* and *"experimentalism"* are nowadays predominant also in computer science, where one has usually spoken of *"empiricism"* before. This demonstrates the growing awareness that the experimenter has nearly total control here, whereas in the social sciences one is very dependent on the selection of questioned subjects.

In his *"wake-up call"* of 1996, Hooker [108] advocates scientific testing instead of pure performance comparisons for heuristic algorithms. His major concern is that only considering efficiency does not promote a deeper understanding of the factors that lead to better or weaker performance, be they test problem properties or parameters. Although comparisons, at least in EC, are not subject to machine speed and code quality—as measurements are usually done in terms of function evaluations rather than milliseconds—Hooker may still be right in suggesting property-based benchmark suites. Many of the employed test sets have coevolved with the algorithms most successfully applied to them and thus are probably biased towards these. Hooker strongly demands concentration on the scientific aspect of experimentation: understanding how and why things work, even at the price of failure. One could argue that Hooker outlines a vision of *experimental analysis*, even if not explicitly termed as such.

Johnson [114], himself a theoretician, approves of performing experiments, especially in cases where theoretical results are not easy to come by. In particular, he supports experimental analysis of algorithms that targets understanding strengths and weaknesses of a proposed method in practice. However, he warns of several pitfalls to avoid in order to preserve the meaningfulness of the obtained results. Some important points from his agenda are:

- use of instance testbeds that can support general conclusions,
- use of efficient and effective experimental designs,
- preservation of reproducibility and comparability,
- mention of negative results,
- establishment of well-justified conclusions and search for explanations, and
- informative presentation of data.

Experimental studies on algorithms proceed in four successive steps, as in the natural sciences: task definition, setup of experiment, execution of experiment, and analysis. In principle, this is also what we find in published work on such studies. However, surveys such as the one conducted by Cohen [52] have long since argued that the task definition is often too vague, so analysis and final results are obscure. Moret and Shapiro [153] therefore require explicitly separating the steps, as depicted in Figure 2.1. A similar circular procedure is also suggested by Feitelson [79].

Undoubtedly, it may be helpful to dedicate some of the available resources to exploratory experimentation and search for patterns in the results (e.g., Johnson [114]). This phase can give rise to interesting questions or hypotheses, but afterwards the experimenter must stick to the plan and simply execute what has been fixed beforehand.

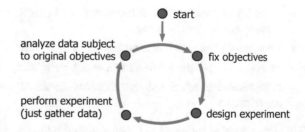

Fig. 2.1 The experimental cycle as outlined by Moret and Shapiro [153]. Changes to investigated questions or experimental design are only allowed when starting a new round.

There are several reasons why putting additional emphasis on explicating the tasks of an experimental study is important:

1. We bind ourselves to a clear task that shall be pursued with a certain steadiness afterwards. This explicit specification enables us to detect deviations easily, which would be impossible otherwise. A clear task is also a precondition for a good experimental setup.
2. Readers of the work are allowed to judge the degree of task fulfillment and importance of findings, resulting in increased scientific transparency.

Moreover, a clear task definition surely supports the third step of the cycle, the data gathering phase. As computer programs are much more volatile than growing plants or machine prototypes, the experimenter is easily tempted to change parts of the setup during the experiment if some patterns emerge from the already available data. When doing so, he is in great danger of losing sight of the original aims and instead preparing answers for questions which have not been posed.

2.2.3 And What About EC?

The previously described self-structurization process of experimentalism in algorithms entailed some movement within the EC community, too. Whitley et al. [241] mainly address the difficulty of selecting suitable benchmark problems. Like Hooker [108], they argue in favor of property-based benchmark suites and in particular demand for test problems which are (one or a combination of) multimodal, nonlinear, nonseparable, nonsymmetric, and scalable. These problems may be constructed by combining several simple functions into one. Additionally, they call attention to evaluation cost changes within one problem, e.g., by a strong dependency on the number of variables.

In spite of these suggestions, the years have seen many so-called *horse race* papers[1], Eiben and Jelasity [71] have reported on the experimental EC and the reasons for their discontent with current practice. Put positively, their main demands were:

- assembly of concrete research questions and concrete answers to these, no claims that are not backed by tests;
- selection of test problems and instances motivated by these questions;
- utilization of proper performance measures relating to the questions; and
- reproducibility, which requires all necessary details and/or source code to repeat experiments.

This line was also followed in Eiben and Smith [74], in some more detail. Partly in response to these demands, Bartz-Beielstein [16] set up a structured experimentation procedure that concentrated on algorithm parameter investigation, based on modern statistical techniques. The current state of experimental methodology development in EC and related areas is summarized in [18].

Concerning benchmark suites, the last few years have seen at least two notable developments, both connected to "*competitions*" between algorithms held in connection with conferences. The first joint approach for the CEC 2005 [224] resulted in a property-based collection of 25 test functions, including composite (hybrid) problems. These became widely used in the literature and the competition was regularly repeated during the following years, mostly with a slightly different focus (e.g., for constrained problems), and the employed problem sets also developed with the requirements. The newest version of this branch of benchmark suites is the one used

---

[1] Johnson [114] uses this term for papers that aim at showing predominance of one algorithm over (all) others by reporting performance comparisons on (standard) test problems.

at CEC 2013 [130] and consists of a slightly enlarged set of the one from 2005, with some improvements in the experimental setup.

Where the CEC benchmark set focused on the quality achieved after certain numbers of function evaluations, the second notable development, the *black box optimization benchmark* (BBOB), suite puts the emphasis on the *expected runtime* (ERT) for predefined target values. The BBOB set was first established for the GECCO 2009 conference [102, 100], with other issues following in 2010, 2012, and 2013. The set of 24 functions in the noiseless case and 40 functions in the noisy case has not been changed over the years, but the available tools for visualizing and analyzing the data have been successively improved. The 2013 issue is especially targeted for expensive optimization on small budgets. Using the BBOB software, a full report, including tables and pictures, can be generated nearly automatically in minutes. Both benchmark suites are available in different programming languages, thus making them easily accessible.

Other problem collections and structuring approaches include taxonomies, such as suggested by Ashlock, Bryden, and Corns [12] with respect to *graph based evolutionary algorithms* (GBEAs). Taxonomies may also be helpful when applied to algorithms intead of problems, as Calégari et al. [42] demonstrate. However, to the best of our knowledge, none of the taxonomies suggested to date had a large impact.

Recapitulating, we may state that the field currently is in a state of flux, with a lot of progress achieved in the last few years, but with still many problems remaining largely unsolved.

2.2.4 Something Different: The Algorithm Engineering Approach?

While reasoning about experimental research in EC—which is also considered as a collection of metaheuristics—it may be useful to take a look into another related but more theoretically oriented field, namely *algorithm engineering* (AE).

Both approaches, AE and EC, apply theoretical as well as experimental means for achieving scientific progress. However, they do so in different ways, in line with their distinct heritages. Whereas in AE (Cattaneo and Italiano [44], Demetrescu, Finocchi, and Italiano [62]), experimentation is seen as complementary to theoretical advances, the opposite is usually the case in the metaheuristics field, where few applicable theoretical results exist. As already stated by Anderson [9], the theoretical perspective (especially for asymptotic analysis results) resembles the top-down view, the experimental perspective the bottom-up view. He suggests considering simple model-based experiments or estimations as an alternative to putting much effort into full-fledged experimentation.

For metaheuristics, this choice usually does not exist because theoretical analysis proves to be very hard in most cases. Besides the stochastic nature of performance values, it is the desire to obtain concrete statements concerning small input sizes (which is nontrivial for many nonlinear real-valued problems) that renders traditional analytical tools almost useless. Real-world optimization problems mostly contain a fixed and very often small set of input variables, so results on the scalability of an algorithm are not useful.

Furthermore, it is hard to clearly define the goal when assessing a metaheuristics performance. As the name suggests, metaheuristics must cope with many problems left over by the more specialized algorithms. They are by design not well adapted to any single problem class. Rather than refine already established theory, experimentation with metaheuristics must target finding out what exactly they are good for. This often initiates an improvement circuit similar to the one postulated in algorithm engineering, but with the task of adapting a method to a problem rather than analyzing its performance. So, in many cases, there is no alternative to implementing and testing metaheuristics, even if we are not able to predict in advance how they will perform.

The two approaches also differ considerably in performance measuring. On the one hand, assessment of metaheuristics is usually simplified by abstraction from many underlying factors via their black-box complexity (Droste, Jansen, and Wegener [67]). This is useful, but often leads to oversimplifications. On the other hand, performance measuring generally has to take two criteria into account: speed (in terms of black-box evaluations) and solution quality. In real-world scenarios, the available running time is limited, and it is rarely possible to determine the global optimum of a problem. Thus, one has to compromise between the two requirements.

It is clear that experimentation in either approach is not trivial and needs to be carefully planned, as stated by Hooker [108], Johnson [114], and Moret and Shapiro [153]. For metaheuristics as well as for algorithm engineering, the main objective of a well-developed experimental methodology is to support real-world application of the designed methods. This is often done by generalization on the basis of test problems or real-world data. Many of the lurking pitfalls are similar, e.g., biased experimentation, missing reproducibility, or argumentation on the basis of an arbitrarily selected benchmark set. Chimani and Klein [47] give a good introduction to these problems and their solution in the AE area, and McGeoch [145] summarizes problems and methods in a more general context.

Despite all conceptual differences, both approaches face the same problems concerning the issue of experimental methodologies. We claim that the similarities are strong enough to justify comparing these methodologies and assess what they can learn from each other to further a "*science of algorithm testing*" (Demetrescu, Finocchi, and Italiano [62]). Algorithm engineering may profit from metaheuristic approaches to establishing generalizations and to dealing with approximate solutions. In turn, research on metaheuristics may improve integration of theoretical and experimental findings, standardization of test cases based on real data, and consideration of low-level factors (hardware, software).

2.3 Deriving an Experimental Methodology from Sequential Parameter Optimization

What are the main aims and concepts of sequential parameter optimization?
What additional concepts do we need to obtain a complete experimental methodology?

The approach that since has been termed SPO (for *sequential parameter optimization*) was proposed and demonstrated first in Bartz-Beielstein [21]. It is mainly concerned with the investigation of algorithm parameters. However, it is more than that. One could argue that the approach consists of two rather different things on different conceptual levels:

- A basic methodological framework (Table 2.1, S-1 to S-4, and S-11 and S-12) which facilitates target-oriented experimentation with properly embedded statistical techniques. It does so by enforcing explicitness at all levels of experimentation, namely by fixing tasks, algorithms, parameters, and statistical tests that will be used to evaluate the results.
- A model-based parameter tuning technique that determines well-performing algorithm instances and also enables investigating parameter interactions (Table 2.1 and S-5 to S-10).

The two complement one another and their combination leads to a tool that is well suited for heavily parameter-dependent non-deterministic optimization algorithms. As SPO was originally conceived for EAs and related methods, it does not surprise us that these comply with the given characterization.

For all but the so-called "*parameterless*" evolutionary algorithms, setting proper algorithm parameters without the help of an at least partly automated tuning procedure—taken seriously—is a laborious process; it has often to be performed manually, one parameter after another. Even worse, many experimental studies of the past did not pay much attention to parameter setting at all. Or they employed default parameter values taken from previous work, without even addressing the question about whether these would work well on the treated test problems.

However, several algorithm parameter investigations utilizing SPO have shown that performance may increase dramatically if appropriate parameter values are provided, e.g., Bartz-Beielstein, Preuss and Rudolph [25], and Stoean et al. [218]. In some cases, the speed of otherwise equal algorithms was increased by factors of between 10 and 100, measured in the number of evaluations needed to reach the optimum with predefined accuracy. It seems likely that criticism of current EC experimental methodology as stated by Eiben and Jelasity [71] still rather underestimates the effect of algorithm parametrization on performance. Consequently, the level of ambiguity induced in past experimental studies by not emphasizing parameter adjustment must have been enormous, rendering a lot of this work practically useless. One of the primary aims of SPO is to reduce this level and thus to enhance reliability of experimental investigations.

Improving reliability by diminishing the threat posed by arbitrariness (see Sect. 2.1) nonetheless only enables, and does not guarantee progress with respect to the fundamental driving force behind SPO: *to advance our understanding of evolutionary and other stochastic optimization algorithms by means of experimentation.* Stated differently, one could argue that applying parameter tuning techniques puts us in the position of obtaining meaningful answers if we pose the right questions, but it does not help in finding these.

Obviously, a tuning-based experimental methodology does not remove all degrees of freedom. But despite that, it presumably allows for more reliable experimental results if the choices made lead to settings that eventually resemble targeted real-world deployment scenarios. Tools such as SPO cannot replace the creative experimenter, but they may be able to support him well. Deriving meaningful experimental results still remains a difficult task. As stated by Schwefel: *"The reader may be warned: Performing a good experiment is as demanding as proving a new theorem"* (Foreword to Bartz-Beielstein [16])

Table 2.1 *SPO consists of three phases on two conceptual levels. Phase I (top): experiment construction; phase II (middle): tuning core; and phase III (bottom): Result evaluation. Phases I and III build the basic methodological framework, and phase II is especially targeted at highly parameter-dependent optimization algorithms such as EAs. It may be replaced with a single step run of the designed experiment if no tuning shall be applied.*

Step Action
(S-1) Pre-experimental planning
(S-2) Scientific claim
(S-3) Statistical hypothesis

(S-4) Specification of the
    (a) optimization problem   (b) constraints
    (c) initialization method   (d) termination method
    (e) algorithm (important factors) (f) initial experimental design
    (g) performance measure

(S-5) Experimentation
(S-6) Statistical modeling of data and prediction
(S-7) Evaluation and visualization
(S-8) Optimization
(S-9) Termination: If the obtained solution is good enough, or the maximum number of iterations has been reached, go to step (S-11)
(S-10) Design update and go to step (S-5)
(S-11) Rejection/acceptance of the statistical hypothesis
(S-12) Objective interpretation of the results from step (S-11)

2.3.1 The Basic Methodological Framework

To increase the expressiveness of experimental results concerning evolutionary algorithms, the SPO framework employs three basic measures. It enforces making the
aim of an experiment (research question) explicit, it integrates statistical techniques
to transform measurements into statements, and it identifies the basic components
utilized for an experiment, which are then arranged into a structured process. The
established procedure is inspired by the lessons learned from experimental analysis,
as summarized in Sect. 2.2.2.

In Figure 2.2, we give an overview of the whole experimental procedure and its
different layers. The concept behind employing a scientific and a statistical context is
to deliberately disclose the necessary transfers between research question, scientific
claim, and statistical hypothesis on the one hand and the outcome of statistical tools
and scientific results on the other hand. A research question that originates in the
scientific context must be translated into a statistical hypothesis to be verifiable.
Setting up a scientific claim as an intermediary step facilitates this context switch; a
scientific claim already is a statement that is falsifiable in principle. Yet it must be
expressed in terms of a statistical hypothesis to enable our approving or rejecting it
in a defined and reproducible way.

Once the measured outcome of an experiment becomes available, it can be treated by
means of the previously determined statistical tools. The obtained statistical data—
e.g., test results, significance levels, and confidence intervals—have to be retranslated
into the scientific context to finally enable our judging the scientific claim and thereby
answering the original research question. Note that traversing the layers becomes
inevitable as soon as any statistical tool gets involved. It is not new in the sense that
SPO would have invented it; rather, it makes the existing layers apparent.

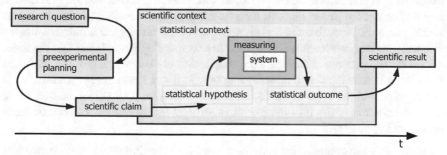

Fig. 2.2 Steps and layers of an experiment within SPO's methodological framework. First, experimentation leads to a scientific claim, which is translated into a statistical hypothesis. This is tested
by means of measuring a system; test results are retranslated into the scientific context.

2.3.1.1 Research Questions, Claims, and Hypotheses

At the beginning of an experimental investigation, it is often not completely clear what its exact task will be. Usually, one starts with a rather unspecific research goal[2] or question that consists of little more than a set of applications under consideration (e.g., several algorithms on a fixed problem set) and a general idea of what one expects to find, based on the experience of the experimenter or on opinions expressed in the relevant literature. In this situation, it is common to take some preliminary measurements. Within SPO, this step is called *pre-experimental planning* and its course is not restricted in any way. On the contrary, its character is explorative and eventually leads to one or multiple scientific claims.

One could argue that this phase is not necessary—it uses up a lot of precious time and effort. However, we recommend taking it seriously because it serves several purposes. At first, the expectation of the experimenter concerning the considered applications may prove wrong. This would require severe changes to the designated experimental setup. For example, an algorithm may dominate all others so clearly that changing the problem or algorithm set in focus (e.g., by taking in problems which evoke different behavior or algorithms that are more similar to the dominating one) gets inevitable. Closely related to this situation are the floor and ceiling effects. If all obtained results are near either the worst or the best possible outcome (as sometimes happens for badly chosen success rate performance criteria), the experiment would be useless. Generally, it appears to be a good idea to verify that the considered performance criteria are able to generate meaningful data. Lastly, one may encounter various forms of surprises; e.g., an unknown effect may show up that must be taken into account when fixing the setup of the main experiment.

Once it becomes clear that the interplay of the scheduled experimental constituents (algorithms, problems, performance criteria) works and is capable of generating useful data, it gets time to determine a concrete scientific claim. To find evidence in favor of or against this claim will then be the task of the main experiment. It was already stated before that the algorithms considered herein are nondeterministic; although it is common to accumulate their results into single numbers (e.g., the mean value) for the purpose of comparison, they rather are treated as random variables that obey a certain distribution. Accepting or rejecting a claim based on such data is not possible without applying statistical tools. Therefore, it simplifies matters if the scientific claim is easily transferable into statistical terms, resulting in one or many statistical hypotheses.

In addition to a hypothesis that is derived from the scientific claim, a test procedure and all its parameters should be fixed now. To exclude any possible feedback from the obtained results into the test procedure, this must be completed before the main experiment starts. By doing so we avoid any type of 'biased interpretation' that would result from setting up the test conditions after knowing the outcome of an experiment.

---

[2] Bartz-Beielstein [16] lists four standard goals: discovery, comparison, conjecture, quality.

2.3.1.2 Hypothesis Testing

In case the output of the investigated applications is deterministic, it is straightforward to determine if it fulfills a mathematically formulated condition—a hypothesis. However, results of nondeterministic applications are samples of unknown distributions. This complicates such a decision enormously, and, additionally, exposes it to the possibility of error. Without complete knowledge of the underlying distributions, approving or rejecting a hypothesis is always prone to error, which is the more likely the smaller the available sample set is.

The most basic conceivable hypothesis is the one that requires equality of two measured systems, the null hypothesis H_0. The alternative hypothesis H_1 then states that the two systems are different, or, more concretely, how they differ (e.g., the values of one system are *significantly* larger or smaller). Significance is usually expressed in terms of a p-value, which resembles an estimation of the probability for generating the data at hand under the condition that H_0 is true (the systems are equal). This approach comes with at least two problems:

- A p-value does not refer to any probability regardless of whether the null hypothesis is true or false; the significance argument is much more indirect. From stating that it is highly unlikely to produce a received result if a certain condition is true, one concludes that it must be wrong.
- To obtain yes/no decisions from p-values, they have to be compared to predefined *significance levels*. If nothing suggests a special treatment of the results, this level α is often fixed at 5%. However, its determination is (a) somewhat arbitrary and (b) also means that at least 1 out of 20 comparisons are wrongly decided because this is the probability of producing a significant sample difference where there is none in the original distribution (type I error). Additionally, in some cases the samples may not allow for a distinction even though the underlying distributions are significantly different (type II error).

As an extension to the concept of significance for lesser sensitivity to arbitrarily chosen parameters, Mayo [140] advocates the use of *severity*. Bartz-Beielstein [17] discusses the application of severity in an EC-related context. It employs metastatistical methods such as the *observed significance levels*, which relate the probability of wrongly rejecting hypotheses to the true differences between the distributions and the number of repetitions. Especially for interpreting test results in light of a scientific claim, defining the largest scientifically unimportant values can improve the adequacy of the test procedure. These allow us to incorporate additional knowledge and thus adjust it to any deployment scenario.

To accomplish hypothesis testing, we have to determine a concrete test method. There, we have to chose between two very different approaches.

- The well-known (Student's) t-Test exists in various forms. They are examples of *parametric* tests because they test against the defining parameter values (e.g., mean, variance) of distributions whose shape they assume as given (usually the normal distribution).

- *Nonparametric* tests do without such strong assumptions, but are often weaker than their parametric counterparts—they need more data to arrive at the same significance level. As an example, we name the Wilcoxon signed-rank test, which resembles the paired t-Test but does not require normally distributed random variables.

Many more test methods of these two types are described in statistical textbooks, e.g., McCabe and Moore [144] and Dalgaard [54]. Note that for multiple best-of-run results of most optimization algorithms, the normality assumption does not hold. Although parametrized tests are often robust against deviations from this assumption, the use of nonparametric tests shall be considered instead. We also discuss resampling approaches that became practicable with the appearance of modern computers. These nonparametric variants utilize extensive sampling from the available data to increase accuracy. The use of bootstrap- and permutation-based hypothesis tests is described in Good [94].

2.3.1.3 The Setup: Design and Terminology

Once claims and hypotheses are fixed, and thus the procedure to apply to the output is defined, it is time to set up the experiment itself. As we want to investigate stochastic algorithms, the performance of which is expected to strongly depend on the chosen parameter setting, it is obvious that running these algorithms several times and computing the mean of the results does not suffice. Whenever multiple factors exist that may influence the outcome, and given that the type of influence they exert is largely unknown, an experiment is 'designed' carefully. The chosen sample points are meant to represent the sampled system as accurately as possible.

General DOE literature (e.g., Mason, Gunst, and Hess [139], and also Montgomery [152]) terms the specifically parametrized system that is probed to generate a measurement *experimental unit*. Works explicitly referring to computer experiments (e.g., Sacks et al. [194], and also Lophaven, Nielsen, and Søndergaard [213]) often use the term *design site* instead, thereby emphasizing the mathematically inclined viewpoint of a location in a large available parameter space which is actually sampled. We adopt this labeling here. The combination of a set of design sites each of which can be sampled concurrently is then termed a *design*.

What are the defining constituents of a design site? In principle, we have to consider all factors that may influence the result. Whereas in classical DOE these are usually split up into the fixed, the controllable, and the non-controllable factors, the last group is only of minor importance in computer experiments. This is due to the freedom to virtually exchange all experimental components without much effort in a computer experiment, as opposed to a laboratory environment. When moving into

Fig. 2.3 Components of an optimization experiment. The algorithm (program) is completed by providing concrete parameter values. The utilized performance measure is usually chosen in accordance with the test problem. It comes into play after the algorithm is terminated and condenses the run data into (mostly) one scalar value.

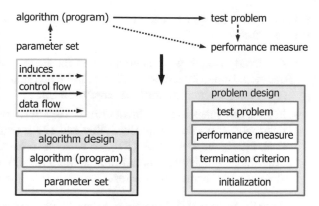

the context of computer experiment-based optimization, another distinction may be more suitable, as suggested by Barr et al. [15]:

- Algorithm factors comprise all 'internal' elements which may influence the behavior of an algorithm: its parameters.[3]
- Problem factors define the application-specific environment of the algorithm, the task it has to cope with. They include start and end conditions, and the employed performance measure(s), as shown in Fig. 2.3.
- Test environment factors encompass all possible sources of influence not covered by the other two. These may originate in computing hardware or software.

If undertaking an experimental study in EC, in most cases one may assume a black-box scenario in which all evaluations are equally costly. Implicitly, one often additionally conjectures that computing these evaluations consumes the largest part of the running time. In this situation, most possible influences on the test environment factors diminish into insignificance and only algorithm and problem factors remain. SPO keeps the two remaining factor groups distinct and describes designs as a combination of algorithm designs and a problem design. The latter is usually supposed to consist of fixed values that represent a single optimization problem (not necessarily one instance). Note that we use "*parameter*" as a synonym for "*factor*", and "*parameter value*" instead of "*factor level*" because it is more common in EC.

2.3.1.4 Reporting Experiments

Surprisingly, despite around 40 years of experimental tradition, in EC a standardized scheme for reporting experimental results never developed. The natural sciences,

---

[3] Note that this may also include alternative algorithm parts as search operators, which may be coded as categorial parameters.

e.g., physics, possess such schemes as de facto standards. Where does this difference originate from?

- As already stated, experimentalism in the natural sciences has a very long tradition. Unlike in computer science, the actual running time of experiments is rather long. Results are thus extremely valuable.
- In computer science, experimentalism is a relatively recent phenomenon. Many important works date from the 1980s and 1990s (e.g., McGeoch [43], Sacks et al. [194], and Barr et al. [15]). In relation to the time needed to set up an experiment, the actual running time is often rather short. This entails a much more volatile character of the results.

Additionally, the impact of nondeterminism on the outcome of EC algorithms may have been underestimated in the past. The result of a comparison between two deterministic computer programs is sufficiently described by reporting its output values. As soon as any stochastic process is involved, much more data has to be provided and taken into account, including at least the recorded performance value samples or suitable statistics derived from them. For the sake of clarity and reproducibility, it is therefore the more important to properly report an experiment's composition and outcome the more it depends on subjective choices (e.g., of performance criteria) and the less exact its result are.

We argue that for scientific readers as well as for writers, a well-defined report structure is beneficial: As with the common overall publication structure (introduction, conclusions, etc.), a standard provides guidelines for readers, what to expect, and where. Writers are steadily reminded to describe the important details needed to understand and possibly replicate their experiments. They are also urged to separate the outcome of fairly objective observing from subjective reasoning. Therefore, we propose organizing the presentation of EC experiments into seven parts, as follows.

ER-1: **Research Question**
Briefly names the matter dealt with, and the (possibly very general) objective, preferably in one sentence. This is used as the report's 'headline'.

ER-2: **Pre-experimental Planning**
Summarizes the first, possibly explorative, program runs, leading to task and setup (ER 3 and 4). Decisions on employed benchmark problems or performance measures shall be taken according to the data collected in preliminary runs. The report on pre-experimental planning shall also include negative results, e.g., modifications to an algorithm that did not work, or a test problem that turned out to be too hard, if they provide new insight.

ER-3: **Task**
Concretizes the question in focus and states the scientific claim and the derived statistical hypotheses to test. Note that one scientific claim may require several, sometimes hundreds, of statistical hypotheses. In the case of a purely explorative study, as with the first test of a new algorithm, statistical

tests may be not applicable. Still, the task should be formulated as precisely as possible.

ER-4: **Setup**
Specifies problem design and algorithm design, including the investigated algorithm, the controllable and the fixed parameters, and the chosen performance measuring. The information provided in this part should be sufficient to replicate an experiment.

ER-5: **Results/Visualization**
Gives raw or produced (filtered) data on the experimental outcome; additionally provides basic visualizations where meaningful.

ER-6: **Observations**
Describes exceptions from the expected, or unusual patterns noticed, without subjective assessment or explanation. As an example, it may be worthwile to look at parameter interactions. Additional visualizations may help clarify what happens.

ER-7: **Discussion**
Makes decisions about the hypotheses specified in part 3, and provides necessarily subjective interpretations for the recorded observations.

This scheme is tightly linked to the 12 steps of experimentation suggested in Bartz-Beielstein [16] and depicted in Table 2.1, but on a slightly more abstract level. The scientific claim and statistical hypothesis are treated together in part ER-3, and the actual execution of the experiment (possibly including parameter tuning) is hidden in part ER-5. In our view, it is especially important to divide parts ER-6 and ER-7, to facilitate different conclusions drawn by others, based on the same results/observations. This distinction into three parts of increasing subjectiveness echoes the suggestions of Barr et al. [15], who distinguish between results, their analysis, and the conclusions drawn by the experimenter. We will employ this report organization for all experiments in this work.

2.3.2 Tuning Methods

The task of detecting appropriate parameter values for concrete instantiations of EAs is not really new; e.g., Grefenstette [96] and Kursawe [121] suggested meta-EA approaches to achieve reasonable settings. However, we may now assess a certain tendency to put more emphasis on experimental analysis. That is, we are nowadays more inclined to understand why specific parameter settings do work on a certain problem while others do not. Many of the recently introduced methods are at least partly oriented towards understanding how the single constituents of an algorithm collaborate to effect a particularly good or bad performance. We attempt a systematized summary of available techniques which have been propagated in the

EC and metaheuristics communities during the last few years. These mostly follow one or two out of three common paths for attaining good parameters.

Meta-Optimization The parameter tuning task itself is treated as an optimization problem, with algorithm parameters as variables (Birattari et al. [34, 33]: racing; Yuan and Gallagher [250]: Meta-EAs combined with racing; ParamILS [111] of Hutter et al.).

Model-Building From sampling the parameter space, a model is instantiated, which enables predicting algorithm performances for yet untested design sites (Ramos et al. [186]: logistic regression; Nannen and Eiben [156]: *relevance estimation and value calibration* (REVAC); Bartz-Beielstein [16]: SPO).

The obtained quality highly depends on chosing a suitable model type for the studied algorithm/problem system. As a newer approach, *Sequential Model-based Algorithm Configuration* (SMAC) [110] by Hutter also copes with categorical parameters by employing random forests as a modeling tool.

Brute Force A thorough exploration of the available parameter space does away with the need for a model, because the data points are so tight that complex forms of interpolation are not necessary. Having real data instead of model answers may help in detecting highly nonlinear parameter behaviors and interactions that would otherwise probably go unnoticed. However, these approaches come with enormous computational costs (Samples, Byom and Daida [196]: parameter sweeps).

In spite of this separation, there are similarities and convergent developments between the different methods, i.e., between racing and SPO.

Racing algorithms, e.g., the F-Race [33], select the best out of a finite number of design sites (continuous parameters may be discretized) by running them several times and deleting the inferior ones by means of statistical tests as soon as significance arises. In each iteration, all remaining configurations are run on one out of a possibly infinite number of problem instances. The advantage is the reduction of the overall number of experiments to determine a single best configuration. We may see it as a 'careful' optimization algorithm because it successively removes the worst performing design sites in order to avoid repeated runs and put more effort into the remaining set of good design points. With the iterated F-Race [35], the authors present an extension of the method that relies more on model building and is thus conceptually very close to SPO.

The REVAC method may also be put into the meta-optimization class as it largely resembles an *estimation of distribution* (EDA) optimization algorithm. However, its focus lies on subsequently updating the internal probabilistic model, which is then used to infer the relevance of and preferable setting for every single parameter.

Subsequent model improvements rapidly lead to detecting the best available parameter settings if the algorithms under scope are deterministic. However, EAs and most other direct search methods are not. Instead of a solid value for any design point, one merely gets an estimation for the desired performance. To make things worse, the obtained distributions are most likely non-normal. Enhancing estimation quality would require us to increase the number of repeats, which can be costly if optimization runs take long. Dealing with non-deterministic algorithms is thus not straightforward but ends up in a trade-off between the number of design sites and the quality of the obtained performance values. Recent results in modelling non-deterministic data [179] however indicate that it may be advantageous in many cases to provide a good spread over the parameter space and completely leave out repeated sampling of one design site.

Note that although the SPO tuning method has proved its usefulness in many cases (e.g., Bartz-Beielstein, Lasarczyk and Preuss [19] on different application scenarios; Bartz-Beielstein, Preuss and Rudolph [25] on comparisons of EA and their hybridization with gradient methods; Preuss and Bartz-Beielstein [172] on parameters for self-adaptation), many of its own recommended parameter values are not based on deep insight into the tuning process but rather on successful ad hoc settings. One of these is the size of the initial design.

2.3.2.1 Initial Designs

Placing the initial population of an EA within the search space or the sample for a first model within a model-based tuning algorithm is a very similar problem. How many points do we sample, and which method do we use for selecting the (design/search) point locations?

In order to answer these questions, we may envision the expected characteristics of the functions we want to sample. For the tuning case, we measure the performance of algorithms in relation to their parameters, which results in non-normally distributed populations that need to be aggregated into single values. The optimization problems themselves cannot be expected to be linear or unimodal.

Consequently, classical "*sparse*" DOE designs are not suitable for capturing the function behavior over the whole search space. So-called "*space-filling*" design methods implement this requirement much better. One of the most popular of these, the *Latin hypercube sampling* (LHS; also LHD for *Latin hypercube design*) was introduced by McKay, Beckman, and Conover [146]. Although many improved schemes have been suggested since then (e.g., orthogonal arrays, Latin supercubes), it is still in widespread use, probably due to its simplicity. With grid sampling and *Monte Carlo sampling*, two straightforward alternative design methods exist. From a geometrical point of view, LHS combines elements of these two with a stratification rule that enables building useful designs with a relatively small number of design points. Without this rule, we arrive at *lattice sampling*.

Simple random (MC) sampling has two important properties, namely (a) every equally sized location in the design space is hit with equal probability, and (b) it produces non-deterministic sequences.[4] However, the first property leads to evenly distributed samples only asymptotically. For small sample sizes, one regularly observes gaps in the design space, lacking points. Thus, the latter property may be forsaken to decrease the variance in point densities. Introducing deterministic random-like sequences also enables reducing the differences between subsequently produced designs. This is the main idea behind *quasi-Monte Carlo* (QMC) sampling and many other related techniques. Giunta, Wojtkiewicz, and Eldred [89] provide a basic and Owen [163] a more technical overview of current sample design methods.

Comparing different sampling schemes may be achieved by using them as integration methods for a concrete function or an unspecified function with certain properties. The smaller the integration error for a given sample size, the better the sampling schema. In this respect, it is known that LHS never performs much worse but often performs better than Monte Carlo sampling on a very general class of functions [163]. Interestingly, [89] demonstrates that in the case of five or more dimensions and reasonable sample sizes ($10^2 \leq n \leq 10^7$), the integration error of MC sampling is smaller than the ones of classical (trapezoidal rule) and QMC methods. Without naming a specific function, theoretical results necessarily remain too general and experimental data too hard to transfer from the utilized test functions. However, a commonly accepted rule of thumb states that the smaller the sample size n and the design space in number of dimensions D, the larger the advantage of complex methods like LHS and QMC. As these numbers are rather small ($D \leq 10, n \leq 1,000$) for most intended EC uses, LHS may be useful for initial models as well as for initial populations.

So how do we choose the initial design/population size? Schonlau, Welch, and Jones [199] recommend $n_{\text{init}} \approx 10 \times D$ for expensive optimization problems treated by means of a model. Recommendations for the parametrization of different stochastic search algorithms may be for much higher values. However, two closely related arguments stand against n_{init} growing too large:

- We require at least n_{init} samples before the optimization/tuning algorithm can actually start to work. During this "*blind flying*" phase, we basically perform random search, which may turn out to be too ineffective.
- As our budget is usually limited, not too much should be invested in this first phase, because it reduces resources for the subsequent phases.

Within the tuning context, an experimental investigation of the trade-off between the initial and the subsequent SPO tuning phases (Bartz-Beielstein and Preuss [22]) revealed that only very general statements are possible without fixing all components (tuning method parameters, optimization algorithm, problem). We presume that for determining the 'ideal' size of an initial population of a stochastic search algorithm, such knowledge would also be necessary.

---

[4] In computer implementations, the cycles of random number generators are usually very long ($t_c \gg 10^{10}$), so the illusion of randomness holds in most cases.

2.4 Parameters, Adaptability, and Experimental Analysis

What is the role of parameters in EAs?
Are we interested in algorithm attributes besides speed and robustness?

In the previous sections, we motivated a structured process of experimentation with a strong emphasis on handling algorithm parameters by tuning them towards the treated problem. However, we now move a level of abstraction upwards and try to conceptually understand what it means to adapt parameters. Particularly, we question the often voiced opinion that uniformly regards all parameters as bad and strives for reducing their number.

What are the advantages and disadvantages of parameterized (optimization) algorithms? On the negative side, we have at least two:

- A large set of parameters makes a method more difficult to handle, because the inexperienced user does not know how to set these correctly.
- Parameter-parameter and parameter-problem interactions may enormously complicate evaluating algorithm performance. To be more specific, it is not entirely clear what kind of performance one obtains if the interactions—and thereby the solution to the parametrization or tuning problem—are not known in advance.

Nevertheless, here we want to further a slightly unorthodox view of parameters, namely of them as handles for modifying algorithms as desired. Simple standard algorithms, e.g., sorting algorithms, are usually considered as being parameterless. However, this is not always the case, as the example of clever quicksort shows. This popular sorting algorithm employs the constant number (3) of elements to determine a median element which is used to partition a set in a *"clever"* way. We could vary the number of elements between 1 and #*elements* and obtain an algorithm that possesses a parameter. However, it is an advantage of simple algorithms with only one parameter that one can often prove that a particular parameter value delivers optimal performance. For evolutionary algorithms running on lots of different optimization problems, the same approach is not feasible. This is partly due to their rather complex stochastic behavior, and also stems from the inherent imprecision of the term *algorithm* as it is often applied to EAs. A concrete algorithm is instantiated only if the two following conditions are met.

1. The problem is clearly specified (e.g., an ES applied to TSP problems is not very specific).
2. All (exogenous) parameter values are fixed.

If this is not the case, one actually deals with an algorithm family, and one usually does so based on theoretical or experimental findings obtained from single algorithm instances. Whereas the first condition leads us to an inevitable dilemma between practical usability and generalizability and thus between the real-world application and the scientific approach, the second one can be dealt with by parameter tuning.

But still, it is not obvious which of the resulting tuned instances a scientific inquiry is based on. The best one obtained by a concrete tuning method with a given time limit? Or a reasonably good (e.g., average) one? De Jong [58] points out that "*getting in the ball park*" by achieving a reasonably well-performing parameter setting is often sufficient for practical purposes, and that doing so may not be too difficult because of a certain robustness of EAs with regard to parameter changes. Nevertheless, we will treat this question again from a different viewpoint in Sect. 2.4.2.

2.4.1 Parameter Tuning or Parameter Control?

Parameter control refers to parameter adaptation during the run of an optimization algorithm, whereas parameter tuning improves the setting before the run is started. These two different mechanisms are the roots of the two subtrees of parameter setting methods in the global taxonomy given by Eiben, Hinterding, and Michalewicz [70]. One may get the impression that they are contradictory and researchers should aim at using parameter control as much as possible. In our view, the picture is somewhat more complex, and the two are rather complementary.

In a production environment, an EA and any alternative stochastic optimization algorithm would be run several times, not once, possibly solving the next problem or trying to solve the same one again. This obliterates the separating time factor. The EA as well as the problem representation will also most likely undergo structural changes during these iterations (new operators, criteria, etc.). These will entail changes in the "*optimal*" parameters, too.

Furthermore, parameter control methods do not necessarily decrease the number of parameters. For example, the 1/5th adaptation rule by Rechenberg [187] provides at least three quantities where there was one—mutation step size—before, namely the time window length used to measure success, the required success rate (1/5), and the rate of change applied. Even if their number remains constant, is it justifiable to expect that parameter tuning has become simpler now? We certainly hope so, but need more evidence to decide.

In principle, every single EA parameter may be (self-)adapted or controlled by means of other mechanisms during runtime. However, concurrent adaptation of many does not seem particularly successful and is probably best limited to two or three at a time. De Jong [58] argues that it would certainly be desirable to hide all parameters by adjusting them inside the algorithm. However, he continues, that based on the experience of the past, this appears to be a very difficult challenge which we cannot hope to overcome soon. We thus cannot simply replace parameter tuning with parameter control.

Nevertheless, there are occasions where the two approaches actually blur into one. If the treated algorithm contains a canonical breakpoint such as the many restarts done within one run of a CMA-ES, it may be parametrized differently, backed up by a surrogate model as also used in SPO-based parameter tuning. At this point, tuning methods are applied during one optimization run, effectively rendering it a parameter control strategy. This has been suggested by Wessing, Preuss, and Rudolph [240], and also by Loshchilov, Schoenauer, and Sebag [132]. These first studies indicate that the approach has a lot of potential for future developments.

2.4.2 Adaptability

For years, experimental research in EC was mainly based on comparisons of standard and newly developed algorithms on a more or less constant set of structurally simple benchmark functions. Interestingly, the standard algorithms evolved slowly, despite the fact that almost all experimental studies of comparisons with new algorithms have *"proven"* that the latter are clearly superior. In the previous sections, we argued that nowadays it is becoming more and more apparent that this inadequacy is to a large extent due to the applied experimental methodology and may at least partly be treated by applying parameter tuning methods.

Almost the same situation appears to be ubiquitous in the practical application of EAs to specific real-world problems. Based on the experience of the algorithm designer, a canonical EA is usually adjusted according to the interpretation of first results, thereby largely following intuitive reasoning and the direction pointed to by simple comparisons. Rough guidelines for the design of problem-targeted operators and representations are provided by the concept of *metric-based EAs* by Droste and Wiesmann [68], but no generally applicable structured method exists to attain a good optimization algorithm for a yet untreated problem. Besides, evolutionary algorithms are so flexible that it seems unreasonable to approach such a method. Parameter tuning may help to a certain extent, but only insofar as (a) there is something to tune, that is, the employed operators are parametrized, and (b) the chosen components of the tuned algorithm are well suited to the problem at hand.

However, availability of fairly automated tuning procedures enables us to look at the adaptation process from the opposite direction: *What does parameter tuning of an optimization algorithm towards one or several problems tell us about the algorithm?* Parameter tuning—or stated differently, parameter optimization—enables us to ask for the best possible performance. But in our case, we are interested not only in peak performance, but also in the way to get to it, in the adaptation process itself. This leads us to the concept of *adaptability* as laid out by Preuss [169].

There are some similarities between this reversed view and the approach of Langdon and Poli [123] to evolve problems which allow for insight into performance properties of different optimization algorithms. But compared to their approach, our focus is wider. We strive to work towards answers to the following questions.

1. Given an algorithm-problem system, what is the peak performance and how is it related to an "*average*" parametrization?
2. How much effort does it require to attain peak performance design points?
3. For which group of similar problems do these adaptability properties prevail?

Unfortunately, it is much easier to pose than to answer such questions, and we do not currently possess a consistent methodology to approach them. However, as these three issues were induced by the everyday use of SPO, we can provide some suggestions, supplemented by first findings of rather provisional character.

2.4.2.1 Tuning Potential

Peak and average performance may be estimated from the results of a tuning process. It surely depends on the employed performance measure how these two values can be aggregated into one, but this quantity may then serve as a measure for the tuning potential of an algorithm-problem system. We thereby assume that the average performance approximately corresponds to the one obtained with default parameter settings, or De Jong's "*ball park*". If no other data is available, it may be determined from a random sample from within the parameter space.

Additionally, we also assume that the best performance obtained from a tuning process is near the achievable maximum. It is clear that this would depend on the utilized tuning method as well as on the time available to make several optimization algorithm runs. Especially for real-world applications, this may be a very restrictive factor. Even if it were not, neither the optimal solution to the parameter tuning problem nor the best achievable performance of the tuned algorithm is usually available, so one has to resort to heuristic rules for stopping the tuning process.

Putting the aforementioned objections aside, and presuming that we have obtained two design points that represent peak and average performance, respectively, we may generate the two corresponding (algorithm performance) samples $\mathbf{y}_p$ and $\mathbf{y}_a$. Under the assumption that the original problem is to minimize, we suggest computing an *empirical tuning potential* (ETP) measure from these samples. We deliberately abstain from using any "*target*" performance as it is not generally known.

$$\text{ETP}(\mathbf{y}_p, \mathbf{y}_a) := \frac{\text{median}(\mathbf{y}_a) - \text{median}(\mathbf{y}_p)}{\text{sqr}(\mathbf{y}_a)} \cdot \frac{\text{median}(\mathbf{y}_a) - \text{median}(\mathbf{y}_p)}{\text{sqr}(\mathbf{y}_p)}. \qquad (2.1)$$

The *sqr* in this formula represents the semi-quartile range, half the distance between the lower and upper quartiles. In comparison to other measures, it has the advantage that it does not depend on a specific distribution type (e.g., normal). It may thus serve as substitute for the standard deviation. The first term of the ETP stands for the relative improvement, the second for the spread reduction. By adding the latter, we take into account that algorithms which mostly reach a near-optimal value (e.g., a self-adaptive ES on the sphere model) usually exhibit a much smaller spread than ones that often get stuck far away from it (e.g., an EA without stepsize adaptation on the same problem). The difference between average and peak performance is given as a multiple of the spread of the distribution of the former. The underlying idea is that the properties of the average performance distribution are utilized to describe how good the improved value is.

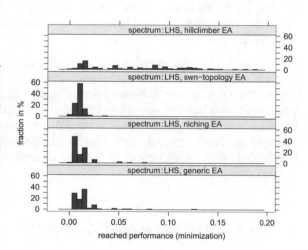

Fig. 2.4 Performance spectrum of four canonical EA variants on a ten random peaks problem as resulting from an LHD of size 100 over the parameter space. From top to bottom: (1+1)-multistart EA, small-world topology CEA (cellular EA), a niching EA as taken from Sect. 6, and a generic EA. The sampling is unbiased towards performance, so that an estimation of the fraction of well-adapted variants is achieved.

2.4.2.2 Tuning Effort

Even more than for the potential, the effort needed to reach maximal performance depends on the utilized tuning method.[5] We argue in favor of two approaches one may choose from, according to available ressources and required accuracy. A rather rough estimation of tuning potential and effort can be obtained from an LHD over the parameter space. As an LHD—in contrast to a more sophisticated tuning method such as SPO—is not biased towards any problem properties, the same design may be utilized for the detection of peak performance, parameter interaction, and tuning effort estimation.

In Fig. 2.4, we provide a simple example of four canonical EA variants on a ten random peaks problem in ten dimensions. Only the three to five parameters regarded as most important (population size, selection pressure, initial step size, learning rate, small-world connection fraction, and restart condition) are checked for, and the obtained samples are used to create a histogram over the achieved performance. From

---

[5] Given that several effective tuning algorithms would be available, all of which detect the best achievable performance. Then it does not seem likely that they all need the same effort to get there.

the frequency recorded for the first performance class, an effort estimation is possible. However, it is very likely that a full-fledged tuning method obtains even better results, so this simple method does not allow for very accurate estimations. Nevertheless, the distribution of samples over the performance spectrum is also informative. Top performance is most easily reached by adapting the swn-topology EA, followed by the multistart EA and the niching EA. Lowering requirements, so that the first two classes in the histogram can be aggregated entails another ordering, led by the niching EA, and followed by the generic EA and the swn-topology EA. The LHD size may be chosen smaller if single runs need more time to complete. In terms of an optimization technique, the LHD method ranges somewhere between random search and grid search.

The second way to obtain estimates for the tuning effort would be to run a more costly but also more accurate tuning method such as SPO. By recording the run number that gave the last performance improvement, an effort estimation is obtained. However, it must be taken into account that due to the nondeterministic nature of the tuned optimization algorithms, SPO runs are also nondeterministic and will most likely result in different peak performances and used efforts (see Preuss and Bartz-Beielstein [172]), which may be aggregated or used to estimate a potential/effort relation.

2.4.2.3 Generalization of Adaptability Properties

Undoubtedly, the generalization aspect is the most interesting as well as the most challenging one. Statements concerning the adaptability of an optimization algorithm to a large group of problems would greatly simplify its application. They would enable making peak performance predictions even for yet untested instances of a problem group. However, we currently lack the means to obtain measurable quantities of similarity between problems. Rather, one is often forced to argue based on phenomenological differences and similarities, which are hard to quantify.

Nevertheless, for two concrete problem classes, the adaptability properties of algorithms may be compared and we may conversely make conclusions about their similarity towards an algorithm from the degree of similarity found in the two systems' tuning behaviors. As an example, we subject a second problem to the LHD (Fig. 2.5) already applied to the first one in Fig. 2.4. They only differ in the number of optima: ten and 100 for the first and the second problem, respectively. In both cases, we operate on problem classes and not on single instances, as the utilized instance set is randomly generated. By employing common random numbers for the generation process, we ensure that all design points are tested under the same conditions.

When comparing the obtained performance spectra, we at first observe that they are very similar. The multistart EA is still the most difficult to parametrize, although adapting it to reach a good performance level is also possible for the 100 peak case. There are, however, some subtle differences. The results for the other three

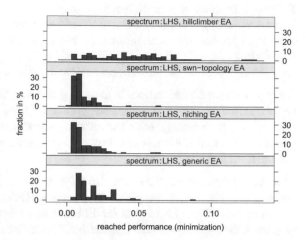

Fig. 2.5 Performance spectrum of the same four canonical EA variants as in Fig. 2.4 on a 100 random peaks problem, again generated from an LHD of size 100 over the parameter space. The Y axis is rescaled as the maximum fractions found in each group are much smaller here. For the swn-topology EA and the niching EA, the fraction of the best performing groups is now much larger than for the naïve EA.

algorithms are stretched over a larger interval, and the naïve EA does not reach the top performance level any more. On this problem, the swn-topology EA and the niching EA are clearly advantageous, as they exhibit better peak performances and larger fractions of well-performing design points, meaning that they are easier to configure. The potential measure of Eq. 2.1 is not applied here, as the fitness of the global optimum is known in advance ($f^{*G} = 0.0$) and we are interested in comparing the absolute performance values.

2.4.2.4 Facing the Future

The previous sections have provided arguments in favor of further investigating the adaptability properties of algorithms and have delivered basic measures and tools. Optimization algorithms will still have to be adapted for concrete applications in the future unless one is found that performs optimally on all problems that will ever be of some importance. We expect that the next few years will see many "*control experiments*" which are undertaken to verify the existing experimental results on the most important EC algorithms, this time taking parameter tuning into account. While for real-world applications simple tuning techniques are likely to dominate, the time constraints for scientific inquiries are usually not that tight. Here, a deliberate decision is necessary to select one or multiple points on the performance spectrum in order to make a fair comparison between different algorithms. Thus, the adaptability viewpoint may have a certain role as it enables integrating the results of a tuning process.

Assessing one algorithm as clearly better than another because its highly interacting parameters have been adapted with a lot of effort whereas the parameters of the other can be handled with much less effort is simply unfair: the two are incomparable.

This make us think of a multiobjective-based view on algorithm performance and in the concrete case requires fixing concrete environmental conditions, taking the adaptation process into account as well. For example, many real-world problems enforce low adaptation times and will therefore favor optimization algorithms with good tuning potential but small necessary tuning effort.

We hope that our considerations help us stop thinking about parameters only in the negative. On the contrary, in approaching practical problems it may be advantageous to insert new parameters into an optimization algorithm. Adding a new, specialized operator does structurally resemble this and is already everyday practice when fitting an algorithm to a new problem.

At an even higher level, one could also think of exchanging algorithms during the optimization process, as suggested by Drepper, Heckler, and Schwefel [66]. The reasoning behind that idea is that for different search space regions with dissimilar properties, distinct methods may be suitable. Given that human expertise is available, change decisions shall be made interactively. Where this is not the case, one may still try out the approach by merging different algorithms into a hybrid one, possibly adapting it to the problem by parametrizing the transitions.

Chapter 3
Groundwork for Niching

Here we establish a suitable definition of niching in evolutionary computation and approach the question of the potential of niching methods in optimization.

3.1 Niching and Speciation in Nature

What is the current understanding of the processes of niching and speciation in evolutionary biology? To what extent do natural processes resemble the situation in optimization?

Importing concepts from evolutionary biology (ecology), which undoubtedly is the origin of the general idea of niching for EAs, appears as promising as it is challenging. Over the years—since niching was first introduced into EC—biologists' view of the phenomenon has also evolved. They now tend to view separation into niches as a process affected living beings actively take part in, also treated as *niche construction* (Odling-Smee et al. [160]). Whereas individuals in canonical EAs are merely collections of values without a *"life of their own"*, living beings act on highly dynamic fitness landscapes and must pursue several objectives (e.g., food and reproduction). Instead of dealing with points in a high-dimensional search space with some internal parameters attached, evolution in nature deals with highly complex and interacting processes, which are only understood partly. We know now that the regulatory system encoded in the part of the human genome that was formerly indicated as *"junk"* has enormous effects on the function of the *"normal"* genes (e.g., Djebali et al. [64]), but it will most likely take a long time before these effects are decoded sufficiently.

The related problem of speciation—the term species often denotes separate subpopulations in niching EAs—currently is one of the most progressive research topics in evolutionary biology, with Mayr's reproductively isolated populations [141, 142] and the allopatric (geographic) speciation mechanism as predominant concepts.

Although these two can be (and are) adapted for use in EAs, biologists still have not reached a consensus concerning all issues raised by the concept of speciation, and are thus not able to provide a simple recipe for arguing with in the EC field. The state of the speciation debate in biology is summarized in Coyne and Orr [53]. We do not go into detail about the biological concepts and their use in EAs, as this has been commendably done by Shir [208] already. There is no doubt that concerning the use of separate populations (which we regard as the core property of niching methods), biology provides a rich source of ideas to transfer into EC. However, there is no guarantee that a mechanism that is found in nature also performs well in an optimization algorithm. This has to be checked by theoretical and/or experimental investigations. Arguing with biological concepts can get adventurous when these concepts themselves are challenged. This also happens to basic beliefs in evolutionary biology, such as the one that older clades ("*branches*" in the "*tree of life*") usually contain more different species throughout their existence than younger ones, which is seemingly untrue at least for the eukaryotes[1] as found by Rabosky, Slater, and Alfaro [185].

As an example of a controversially discussed yet unsolved question, we consider the formation and maintenance of sexual reproduction. This issue is usually dismissed in EA research in favor of asexual populations,[2] for which in turn no widely accepted speciation concept exists in biology. In consequence, biological terms shall be used with care when applied to niching EAs to prevent conceiving meanings where there are only metaphors. Nevertheless, the biological concepts undoubtedly provide an important source of inspiration for niching algorithms, especially for the dynamic case of time-dependent optima; natural evolution with its rich interactions between species can hardly be regarded as static. However, in this work, we will focus solely on static optimization, as even this probably much simpler problem is currently not satisfactorily resolved.

3.2 Niching Definitions in Evolutionary Computation

What are the purpose and the characteristic properties of a niching method in EC, irrespective of niching concepts in evolutionary biology?

Niching in evolutionary algorithms appears to be a heterogenous collection of techniques applied to enhance their ability to cope with multimodal objective functions by implementing some form of parallelization, either in terms of search space or time, or both. Does it contain all EA variants suggested for multimodal optimization? Surely not. But to state what exactly distinguishes niching approaches from other ones seems non-trivial, as—despite existing, partly contradictory definitions—the

---

[1] Eukaryotes are life-forms whose cells possess nuclei.

[2] There are some notable exceptions that feature sexual populations, e.g., for approaching optima on the borders of the feasible search space, as suggested by Kramer and Schwefel [119].

evolutionary computation community apparently does not yet possess a unified taxonomic view on multimodal evolutionary optimization. Much effort has been put into fitting algorithms for solving unimodal problems, thus acting as local search method, which is easier to assess experimentally and is also more theoretically tractable. It is our aim to contribute to a movement of developing optimization algorithms for multimodal problems further by investigating what niching actually is and what it can do to improve evolutionary algorithms.

Out of the large set of publications dealing with niching or similar techniques in EC (e.g., De Jong [56] and Goldberg [91] as some of the earliest), we select only two opinions to show where to locate possible disagreements in defining niching. Mahfoud gives the following functional specification of niching methods in an optimization context:

"The litmus test for a niching method, therefore, will be whether it possesses the capability to find multiple, final solutions within a reasonable amount of time, and to maintain them for an extended period of time."

(Mahfoud [137], page 61)

He additionally states that the multiple solutions correspond to multiple local optimizers. Beyer et al. include the process of separation, too. However, they also add diversity maintenance in their definition:

"Niching—process of separation of individuals according to their states in the search space or maintenance of diversity by appropriate techniques, e.g. local population models, fitness sharing, or distributed EA."

(Beyer et al. [30])

Whenever speaking of niches in EAs for static black box optimization, authors usually identify these with basins of attraction, at least for real-valued optimization. As Mahfoud points out, diversity maintenance is related to niching but must not be pursued too rigorously because it enables but does not guarantee finding many basins, depending on the basin distribution within the search space. In this sense, combining parts of both specifications, referring to basins of attraction, and leaving out diversity maintenance leads us to the following definition:

"Niching in EAs is a two-step procedure that a) concurrently or subsequently distributes individuals onto distinct basins of attraction and b) facilitates approximation of the corresponding (local) optimizers."

(Preuss [168])

Undoubtedly, all EAs have local search capabilities. Therefore, it must be the process of locating *distinct* basins that induces difficulties and requires experimentation with many EA variants to establish a good niching process. Note that the two tasks of evolutionary methods, local and global search, are often referred to as exploitation and exploration (see, e.g., Eiben and Schippers [72] for an overview of possible meanings of these terms). It seems to be a common opinion that these two are contradictory and that an algorithm can do either one or the other well. However, we argue that there are already satisfactory solutions for the local search (exploitation) problem. This gets clear from the results of the *Black Box Optimization Benchmark* (BBOB) competitions held at the *Genetic and Evolutionary Computation Conference* (GECCO) in 2009, 2010, and 2012. The most successful algorithms were restart variants of traditional (e.g., BFGS), direct (e.g., Nelder-Mead) or evolutionary (e.g., CMA-ES) optimization methods.

This has become possible because the current stopping conditions are much more clever than what was applied in the last century and it is thus not necessary any more to do lots of useless evaluations when the pursued local optimum has already been hit. The stopping conditions of the CMA-ES (Auger and Hansen [14] and Hansen [98]) are especially well developed and in principle applicable to other algorithms as well. Whenever we are in a search space area that seems to have potential for further improvement and the chance is high that we do not end up in the same local optimum, it would be counterproductive to stop a local search. Thus, in our opinion the question is rather not when to exploit and when to explore but how, and especially *where*. This last aspect has often been termed *diversity maintenance* in earlier works. Nowadays, the term may be misleading as many (in particular, restart) algorithms do not possess a single top-level population anymore. Nevertheless, diversity or the question of where to place the next local optimization attempt are of course still of utmost interest for the success on multimodal problems.

In accordance with the explicit/implicit diversity maintenance scheme suggested by Eiben and Smith [74], we further partition niching EAs into two groups, performing explicit or implicit *basin identification*. Explicit basin identification methods— detecting the basin of each individual—divide the individuals into subpopulations, according to their basins. Even uncertain knowledge about the individual-to-basin relation may, however, be used also for learning about the structure of an optimization problem, e.g., one can do size and location estimations of the basins.

Note that for basin identification, we only require that, for two points x_1 and x_2 in the search space, an oracle tell us whether they belong to the same basin. In the ideal case, the oracle provides us with perfect information. However, for all non-trivial optimization problems, we can expect only imperfect results, such that the probability $p_{BI}(x_1, x_2)$ of correctly identifying two points as residing in the same basin is less than 1.

Whether basin identification is targeted at dividing a population into clusters resembling basins of attraction, and thus uses only current information, one could also collect all search points obtained during optimization (or gathered from other sources) into an archive and derive a spatial characterization of the known basins. With this at hand, we can decide for every new search point if it belongs to a basin we have already explored. Although this second oracle is very similar to the one for basin identification, we term this functionality *basin recognition*; it may be seen as the offline variant of basin identification. The probability that a point is rightfully detected as being located in an already known (visited) basin will be denoted by $p_{BR}(x_1)$ in the following. Note that this is a classification problem with two classes (known and unknown) and results fall in four cases:

- true positive: the basin is already known and the oracle answers yes
- false positive: the basin is not known but the answer is yes
- true negative: the basin is not known and the answer is no
- false negative: the basin is known but the answer is no

We assume that the number of false positives (type I error) can be kept very low as it means that a basin has never been visited but still the oracle considers it known. However, it does so without any supporting data it could base this decision on. When designing a method implementing the oracle, one should be aware that false positives block a basin from being explored, and thus it should be prevented from answering yes when there is no satisfactory information for this conclusion. A low fraction of false positives means that the ratio of true negatives will be very high. Nevertheless, we can expect a number of false negatives where the basin recognition method fails to recognize that a basin is known. The main task for the oracle is thus to distinguish true positives from false negatives. This capability is known as sensitivity (3.1), and this is what is meant when we speak of probabilities p_{BR} and p_{BI} (a very similar argument can be applied to basin identification).

$$\text{sensitivity} := \frac{\sum \text{true positives}}{\sum \text{true positives} + \sum \text{false negatives}} \ . \tag{3.1}$$

A basin identification method with high sensitivity enables setting up a basin recognition method by application to the point pairs. The reasoning behind this differentiation will get clearer in the rather theoretical treatment of the niching process in the following section, Sect. 3.3.

3.3 Niching Versus Repeated Local Search

Under what conditions can parallelized search (niching) be more effective than (blind) repeated local search methods?

As the stream of new methods suggested for global optimization—within or outside of EC—does not cease, one may ask what the motivation behind designing new niching EAs is. The seemingly underlying yet unreached goal is to convincingly beat one of the simplest algorithms for multimodal objective functions, the iterated/parallelized hillclimber/local search. In the spirit of the NFL, this task appears adventurous when optimizing general multimodal problems, but it may be possible for problem classes exhibiting certain exploitable properties.

In the following, our main task is to gather evidence in favor of or against the prevalent belief (in EC) that niching EAs can outperform (blind) repeated local search algorithms that rely on randomly generated start locations. Note that this is an *existential* precondition for designing new niching EAs, as these are usually algorithmically much more complex. We thus complete the second step prior to the first and simply assume the existence of efficient basin identification and/or basin recognition methods for population-based optimization methods. These would enable deciding if any two individuals are located in the same basin and if a basin has been visited before. Thus we investigate the following question:

> Given that basin identification and/or basin recognition works, how much faster can a niching EA be in terms of a *redundancy factor* (measuring superfluous local searches; see Beasley et al. [26]) than (blind) repeated local search algorithms?

Note that instead of using a blind repeated local search, one could also attempt to use findings obtained in previous local searches when starting the next one. This concept is known as *iterated local search* (ILS; see Lourenco et al. [133]). Certainly, an ILS method could also benefit from basin identification/recognition methods.

In the following, we set up a simple niching model that employs a small number of properties in order to characterize the hardness of an optimization problem.

3.3.1 A Simple Niching Model

Modeling the behavior of a real niching EA on an idealized multimodal objective function still bears enormous complexity. The whole local search process in the detected basins must be considered, and is likely to heavily depend on algorithm and problem parameters. Hence, for our niching model EA, we cut this process out by choosing the number of local searches as the unit of measurement. As already stated, we assume that basin identification and basin recognition are done by specific, abstract methods which are fast and have fixed accuracy probabilities $p_{BI}(\mathbf{x}_1, \mathbf{x}_2)$ and $p_{BR}(\mathbf{x}_1)$. For any real application, both probabilities will most certainly be smaller than 1. However, they should clearly be larger than 0 for achieving an advantage over ILS methods. Note that comparing with a completely randomized oracle does not make much sense as the underlying problem that must be solved before providing a positive answer is to determining if some points $\mathbf{x}_1$ and $\mathbf{x}_2$ reside in the same of possibly many basins; randomized methods cannot be expected to answer this question with high sensitivity. We will discuss how to actually perform basin identification/recognition later in Chapters 4 and 5.

Another objection to only counting local searches is that it may not be possible to detect the basin of an individual as soon as it enters it. Thus, the implied advantage of an ideal niching EA that consists of breaking unnecessary local searches at the start may not be realizable in full. But, unless other techniques are applied to reduce the optimization effort (e.g., utilization of attained information to speed up subsequent local searches), *no* niching EA can be faster in terms of local searches (on average) than the simulation by means of the niching model—we obtain an estimation for a lower bound.

In Sect. 1.1.5, some properties of basins already have been established; these may serve as a starting point for our model. Note that some of these (such as dimensionality) are not important for the model as our search space is a set of basins and the number of local searches is the basic unit of counting. The curse of dimensionality has to be dealt with by the basin identification method. To keep things simple, a possibly existing basin level structure (e.g., a funnel) is not taken into account. Thus, we arrive at four properties defining a situation during an optimization run in our niching model:

- number of basins b,
- basin size contrast (the volume of the largest basin divided by the one of the smallest),
- number of basins c covered by a certain algorithm simultaneously,
- basin identification and recognition probabilities, $p_{BI}(\mathbf{x}_1, \mathbf{x}_2)$ and $p_{BR}(\mathbf{x}_1)$, respectively (the $\mathbf{x}_i$ stand for search points)

Note that we consider c as constant to keep the model simple. It will rarely be the case that the same number of basins is identified for any real procedure and problem, so c is actually a random variable.

Algorithm 1: Niching model algorithm

---

1 **repeat**
2   randomly initialize a number $\geq c$ of individuals on c of b basins;
3   basin identification: match individuals to basins with accuracy $p_{BI}(\mathbf{x}_1, \mathbf{x}_2)$;
4   select one individual per basin $= c$ individuals;
5   **forall the** c *individuals* **do**
6   **if** *basin of c recognized (with accuracy $p_{BR}(\mathbf{x}_1)$)* **then**
7   perform local search on selected individual;
8 **until** *stopped externally (all basins visited)*;

Without basin identification, one is thrown back to iterated/parallelized local searches for which the required effort is known [26]. Covering the whole basin set with randomly initialized local searches results in a number of superfluous local searches as already explored local optima are visited again (for a repeated local search, the distribution of local searches to optima is blind and depends on the size of the basins). This is measured by the redundancy factor R (from [26]). In our context, it describes the situation of equal basin sizes:

$$R = \sum_{i=1}^{b} \frac{1}{i} \overset{b \gg 3}{\approx} \gamma + \ln b, \qquad (3.2)$$

where $\gamma \approx 0.577$ is the Euler-Mascheroni constant. For entering each of the b basins at least once, $R \times b$ local searches are necessary on average. We can conclude from

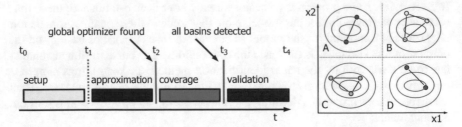

Fig. 3.1 Left: Four phases of a heuristic optimization process. We are interested in detecting t_2 and t_3. For reasons of simplicity, we presume that time spans are given in numbers of local searches. However, other measurements are possible. Right: Niching model EA population after initialization and basin identification. Individuals residing in one basin are connected by lines.

the formula that the relative search overhead shrinks (as the logarithm grows much slower than a linear factor) with the number of basins, so niching methods will be able to outperform restart methods only if b is not too large. On highly multimodal functions without any exploitable basin-level structure, they will presumably not be faster. However, if some basin properties, e.g., size, prevent entering the important basins easily, it should pay off to use a niching method. The real-world problem investigated in [175] provides a simple example: The search space consists of a small number of basins, a very large one providing the global optimum (which corresponds to an unwanted, physically infeasible solution), and several tiny ones of which one contains the (local) optimum that was looked for. Note that the global optimizer had been known in advance, which helped enormously in locating the small one.

Instead of conducting single local searches, a niching algorithm performs several coordinated local searches. In our model, we repeatedly start with a randomly initialized set of individuals and undertake only necessary local searches until all basins have been visited (Algorithm 1). The number of *covered* basins c is not a parameter but a measured size. When distributing a number of individuals, we do not know how many basins we get. This is meant to simulate the situation often encountered in the starting phase of a metaheuristic algorithm: the population size is predetermined. We also do not specify how the local searches are implemented; they may be realized by mating restrictions, or separate populations, or embedded helper methods. Note that basin identification only needs to detect if individuals are located in the same basin; it is not required to properly recognize each basin as such (Figure 3.1, right). Basin recognition only needs to detect if an individual belongs to a basin on which a local search has already been conducted.

What kind of performance data do niching model EA simulation runs deliver? Figure 3.1 (left) displays the phases of any heuristic optimization algorithm in terms of basin detection. During setup, the algorithm is prepared and started and yields the first result in time t_1 (after the first loop we have obtained a set of local optima of which we choose the best one). We assume that time is measured in the number of local searches here, although other measurements would be eligible, of course.

The following approximation phase lasts until the global optimizer is hit for the first time at t_2. It shall be noted that, especially in real-world applications, this point is often not reached because evaluations may be too costly. The coverage phase is needed to visit each basin at least once and ends at t_3. Unless the number of basins is known in advance, it seems almost impossible to determine t_3 from inside an optimization algorithm when it actually occurs. It is up to the user to stop it when no new information can be obtained from running further (t_4). In the case of the niching model EA, t_2 and t_3 are measurable because the basin set is known.

Note that the redundancy factor stated in Equation 3.2 is equivalent to t_3, divided by the number of basins b. We would like to emphasize that it thus does not refer to the expected first hitting time (when the global optimum actually is detected) but to the end of the coverage phase (when all optima are known). In fact, as we shall see, t_2 and t_3 behave very differently. Why should one be interested in t_3, if all that is usually measured (e.g., in the BBOB context) is t_2? The most obvious reasons are:

- t_3 resembles the maximum of t_2 we can reasonably expect over several repeats (if the global optimum resides in the last detected basin). By reducing it, we can bring the variance of t_2 down.
- If the basin containing the global optimum is harder to detect than others, coverage gets more and more important and t_2 will grow towards t_3.
- For practical considerations, we may require not only one good solution but many. At t_3 we have obtained as many as possible.
- On an unknown problem, it is usually important to obtain as much problem knowledge as possible in order to shape the considered optimization algorithm (or select another one), following the concept of *exploratory landscape analysis* (ELA).

3.3.2 Computable Results

The aforementioned model was introduced in a simpler form (without basin recognition and probabilities) in [168]. However, under certain conditions, the expected values for t_2 and t_3 can directly be computed instead of being simulated. This is true if we neglect basin recognition and assume equal basin sizes. In this case, the parallel search with basin identification resembles the *coupon collector's problem* (CCP) with group drawings, and Stadje [217] derives a relation for the moments of $Z_n(A,m)$, the expected waiting time for obtaining n elements of the desired set $A, |A| = l$ out of the complete set S by drawing m different coupons at once.

We derive the equation for the expectation (3.3), and in order to adapt it to our context, we write b instead of s, and c, the number of covered basins, instead of m (in the original formula, s stands for the number of elements of the complete set of coupons). The choice of n and A determines if we get t_2 (for $n = l = 1$) or t_3 (for $n = l = b$):

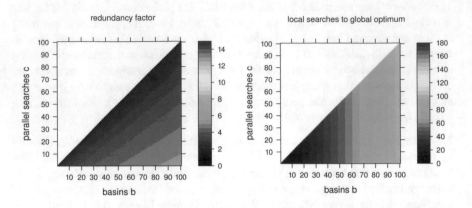

Fig. 3.2 Left: computed expected redundancy factors ($E(t_3)/b$) using equation (3.3); right: local searches needed to locate the global optimum (t_2), derived from the same source. Colors are chosen in order to enable comparison with subsequent figures.

for t_2, only one specific coupon (the basin holding the global optimum) is sought; for t_3, we are required to collect all coupons (basins).

$$E(Z(b,l,n,c)) = \binom{b}{c} \sum_{j=0}^{n-1} (-1)^{n-j+1} \binom{l}{j} \binom{j-j-1}{l-n} \left[\binom{b}{c} - \binom{b-l+j}{c} \right]^{-1}.$$

(3.3)

Figure 3.2 (left) shows the resulting redundancy factors (t_3/b) for up to 100 basins. Note that (3.3) gives the expected waiting time as the number of drawings when, in each drawing, c of b basins are covered. For the number of necessary local searches (t_3) the result must be multiplied by c in the sequential case (one local search after the other is taken out by one processing unit) that is considered here. Depending on the method of measuring, we may slightly overestimate the true number of necessary local searches if $c > 1$. If measuring takes place also within the for loop of Algorithm 1 as is done in the BBOB context, the resulting t_2 and t_3 values are a bit smaller. The error is almost negligible unless c is large and/or b is small. We will obtain a more concrete statement by simulation in Sect. 3.3.3.

The result is accurate for the parallel case given that the basin identification can be done by some central unit and the local searches are distributed, so that (3.3) directly means t_3 (if c processing units are available). Without basin identification, we would do c uncoordinated local searches with $c = 1$ in the same time, such that the redundancy factor again resembles the bottom line in Fig. 3.2 (left).

In order to compute t_2, we set $l = n = 1$, and the equation simplifies to:

$$E(Z(b,c)) = \binom{b}{c} \left[\binom{b}{c} - \binom{b-1}{c} \right]^{-1}. \tag{3.4}$$

By applying $\binom{n}{k} + \binom{n}{k+1} = \binom{n+1}{k+1}$ and $\binom{n}{k} = \frac{n}{k}\binom{n-1}{k-1}$, respectively, we obtain:

$$E(Z(b,c)) = \binom{b}{c} \left[\binom{b-1}{c-1} \right]^{-1} = \frac{b}{c}. \tag{3.5}$$

Thus, the expected value for t_2 (the number of local searches for the global optimum) is b, the number of basins, regardless of the type of search, coordinated or not. Essentially, this means that for equal basin sizes, basin identification helps to lower coverage times, but does not affect t_2, which is at first surprising. However, assuming equal basin sizes is not very realistic with respect to real-world applications. Nevertheless, we are aware that niching methods have some general limitations that prevent them from being cheaper in the number of function evaluations than simple restart methods (b very large and/or very regular basin structure). On the other hand, they shall also not be much slower if the basin identification overhead is not too extreme.

The derived results are summarized in Table 3.1, together with the trivial case of perfect basin recognition, which completely avoids superfluous local searches. Therefore, its expectation for t_2 is $b/2$, and t_3 is always reached after b local searches.

Table 3.1 Expected waiting times for locating the global optimum t_2 and covering all optima t_3 in number of local searches per basin number b, under different BI and BR probabilities (perfect BR and no BI seems very difficult to establish and is therefore left out). Equal basin sizes are assumed.

BI/BR accuracy	$E(t_2)/b$	$E(t_3)/b$
no BI, no BR		1 (3.2) after Beasley [26]
perfect BI, no BR	1 from equation (3.3)	equation (3.3)
perfect BI and BR	0.5	1

Next to the work of Stadje, several other authors dealt with different variations of the CCP. Kobza et al. [116] investigate it under a random number of coupons selected in every drawing. However, the available solutions are difficult to apply as the number of basins identified by a real basin identification algorithm is hardly predictable and strongly depends on the method. For the case of the weighted CCP (resembling unequal basin sizes), Berenbrink and Sauerwald [28] review the available bounds and approximations for obtaining all coupons (t_3), but it currently seems impossible to compute the expected time if c coupons are drawn in each step. Thus, we rely on simulation for the theoretically intractable cases that are of interest in the basin identification context. Namely, we need to know how different basin identification and basin recognition probabilities affect t_2 and t_3 under the equal and unequal basin size assumptions in order to answer the question about when niching is profitable in comparison to repeated local search.

3.3.3 Simulated Results: Equal Basin Sizes

Within this section, equal basin sizes are assumed, that is, the probability of detecting any of the basins is the same. This case does not appear to be very realistic but it is much easier to handle than that of unequal basin sizes and provides a good first impression of what to expect for more complex situations.

3.3.3.1 Experiment 3.1: How Are t_2 and t_3 Affected by Different Accuracies for Basin Identification and Basin Recognition?

Pre-experimental planning

The values in Figure 3.2 provide the baseline for a comparison. Implementing Algorithm 1 is straightforward, and we only need to count the number of local searches that would actually be conducted. p_{BI} is realized by first selecting a number of c distinct basins and then replacing each of these with probability $1 - p_{BI}$ with a random basin number. If a basin has already been seen, then this is recognized with probability p_{BR} and the corresponding local search is not counted. As expected, the standard deviation of the simulation is quite high; thus we perform $1,000$ repeats for every tested configuration to get it down to a level of $\approx 5\%$.

Task

We want to determine if basin identification and basin recognition clearly reduce t_2 and t_3 even for $p_{BI} < 1$ and $p_{BR} < 1$. Defining a concrete criterion is a bit arbitrary, but we will mark an improvement as sufficient if t_2 and t_3 are reduced by at least 10% for $c = \lfloor \frac{b}{2} \rfloor$. Thus, simulation results of t_2 and t_3 for all configurations with $p_{BI} > 0$ will be tested for significance (one-sided) against the original results for $p_{BI} = 0, p_{BR} = 0$ (improvement) and against the same values multiplied by 0.9 (improvement of at least 10%).

Setup

As in Figure 3.2, b and c are set to values from $\{1, 2, ..., 100\}$, but ensuring that c is not larger than b. We try three combinations of p_{BI} and p_{BR} (knowing that p_{BI} gets irrelevant if $p_{BR} = 1$), namely $p_{BI} = 1$, $p_{BR} = 0$ to compare with the theoretical results (which do not account for what happens within one drawing); $p_{BI} = 0.5$, $p_{BR} = 0$ to model an imperfect basin identification; and $p_{BI} = 0.5$, $p_{BR} = 0.5$, to add also an imperfect basin recognition method to the model.

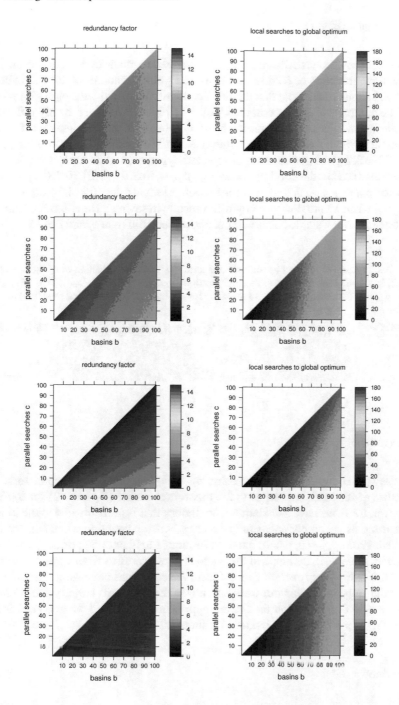

Fig. 3.3 Left: measured redundancy factors (t_3/b), averaged over 1,000 repeats; equal basin sizes. Right: local searches needed to locate the global optimum (t_2), derived from the same simulation. The rows from top to bottom: $p_{BI}, p_{BR} = (0,0), (0.5,0), (1,0), (0.5,0.5)$.

Results/Visualization

Figure 3.3 provides visualizations of the measured redundancy factors and times to t_2. Each combination of b and c results in one point in each levelplot and is based on $1,000$ simulations. Note that the third configuration (row) with $p_{BI} = 1, p_{BR} = 0$ shall be comparable to the computed values of Figure 3.2, with the only difference that during simulation we count the exact number of local searches also if t_2 or t_3 are reached within a drawing. Thus, numbers shall be smaller at least for high values of c. Table 3.2 gives the fractions of rejected hypotheses while comparing t_2 and t_3 values obtained for the configurations $p_{BI}, p_{BR} = (0.5,0), (1,0), (0.5,0.5)$ with the ones for $p_{BI} = p_{BR} = 0$, and also for the latter reduced by 10%, in order to see if applying basin identification/recognition results in savings. Here, b runs from 2 to 100 (testing over a single basin will not show much difference) and $c = \lfloor \frac{b}{2} \rfloor$.

Table 3.2 Fraction of rejected hypotheses concerning significant reduction of t_2 and t_3 over all $b \in \{2,3,...,100\}$ basins for the three configurations $p_{BI}, p_{BR} = (0.5,0), (1,0), (0.5,0.5)$ against $p_{BI} = p_{BR} = 0$ (columns 3 and 4), and the same for the base values multiplied by 0.9 (columns 5 and 6).

p_{BI}	p_{BR}	sig.fraction t_2	sig.fraction t_3	sig.fraction $0.9 \cdot t_2$	sig.fraction $0.9 \cdot t_3$
0.5	0	0.31	0.97	0.00	0.00
1	0	0.98	0.98	0.81	0.96
0.5	0.5	0.99	1.00	0.63	0.99

Observations

It is easy to see the impact of increasing p_{BI} on t_2 and the redundancy factors. The structure of the latter converges to the theoretical model (Figure 3.2) for $p_{BI} \to 1$. However, for t_2 we have a deviation from theory that is probably due to the kind of measuring, as we do not count full drawings but stop counting when t_2 is actually reached. From $p_{BI} = 0$, where c has an influence neither on t_2 nor on t_3 (which is expected, of course), the slope of the contour lines decreases from vertical to around $\frac{1}{2}$ for the redundancy factors and $\frac{1}{4}$ for t_2. It seems that increasing p_{BR} does not affect the slope of the contour lines but moves them towards larger values of b: for $p_{BI} = p_{BR} = 0.5$, reduction factors are approximately halved when comparing to $p_{BI} = 0.5, p_{BR} = 0$, and t_2 is reduced by about 20%

Discussion

The effects of basin identification and basin recognition are obviously additive. Whenever we increase one, we get a reduction of t_2 and t_3. However, under the assumption of equal basin sizes, $p_{BI} < 0.5$ hardly results in a visible advantage. A certain accuracy is needed for the basin identification, and of course c must be larger than 1; the larger, the better. This is also documented in the statistical tests presented in Table 3.2: for $p_{BI} = 0.5$, we obtain a reduced t_3, but the difference in t_2 rarely gets significant. While for the other two configurations there is nearly always a significant difference with the base configuration $p_{BI} = p_{BR} = 0$, this difference is at least 10% only for t_3; for t_2 this is not generally the case. The anomaly for $p_{BI} = p_{BR} = 0.5$ when testing against $0.9 \cdot t_2$ seems to stem from the very low variance for this case. While the three configurations with $p_{BR} = 0.0$ result in standard deviations of approximately $b, 0.9 \cdot b$, and $0.8 \cdot b$ for p_{BI} set to 0, 0.5, and 1.0, the standard deviation for the last case is around $0.6 \cdot b$. If one moves the factor used in the comparison from 0.9 towards 1.0, the fraction of rejected hypotheses quickly converges to 0.99.

Summarizing, we can assess that under the equal basin size assumption we can obtain a speedup for increased $p_{BI} > 0$ and/or $p_{BR} > 0$. Therefore, niching actually does make sense. However, the speedup is not very large unless $p_{BR} \rightarrow 1$. Basin identification alone provides an advantage, but only if c is relatively large in comparison to b (meaning that we have detected several basins at once). For $b \rightarrow \infty$, we cannot expect a measurable advantage of niching methods as we would need a basin identification method that provides us with a number of $c \rightarrow \infty$ identified basins.

3.3.4 Simulated Results: Unequal Basin Sizes

In [168], we have already shown that the situation slightly changes if only moderate size differences of the basins are introduced, even if basin recognition is not considered. Some of the frequently used test problems for global optimization, e.g., the Rastrigin function, feature equally sized basins. However, we cannot expect that this is a general property of real-world problems. But how large should be the differences in basin size considered for investigation?

Note that for a ten-dimensional problem, a difference factor of 1.259 ($\approx \sqrt[10]{10}$) per dimension already leads to a volume difference of 10 to 1. A difference factor of 2 already means a volume difference of $1,024$ to 1. Nevertheless, it makes sense to look at small differences for these reasons:

- It can be expected that basin identifcation/recognition will also get more difficult if size differences get more extreme. Of course, one would need many more samples than are usually available in a starting population of an EA to detect the smaller basins.

- Currently, not much is known about usual basin size differences in real-world problems. We thus restrict our interest to a general idea of the trends that occur due to unequal basin sizes and do not cover all possible cases.

The following experiment shall provide this general idea.

3.3.4.1 Experiment 3.2: How Are t_2 and t_3 Affected If Basin Sizes Are Not Equal?

Task

Same as for Experiment 3.3.3.1, and estimate how large the differences to the equal basin setting are.

Setup

Similar to Experiment 3.3.3.1, but with a maximum basin size difference of $10:1$. Within this range, all basin sizes are uniform randomly determined in every repetition. This means that the sought basin holding the global optimum can get all allowed sizes and is not necessarily small.

Results/Visualization

In the same style as for equal basin sizes, Figure 3.4 shows measured redundancy factors and times to t_2. The same scaling is used for the graphical display. Table 3.3 gives the fractions of rejected hypotheses while comparing t_2 and t_3 values obtained for the configurations $p_{BI}, p_{BR} = (0.5, 0), (1, 0), (0.5, 0.5)$ with the ones for $p_{BI} = p_{BR} = 0$, and also for the latter reduced by 10%. The table can be directly compared to Table 3.2 for equal basin sizes.

Table 3.3 Fraction of rejected hypotheses concerning significant reduction of t_2 and t_3 over $2:100$ basins for the three configurations $p_{BI}, p_{BR} = (0.5, 0), (1, 0), (0.5, 0.5)$ against $p_{BI} = p_{BR} = 0$, and for the base values multiplied by 0.9, for a basin size contrast of $10:1$.

p_{BI}	p_{BR}	sig.fraction t_2	sig.fraction t_3	sig.fraction $0.9 \cdot t_2$	sig.fraction $0.9 \cdot t_3$
0.5	0	0.37	0.97	0.02	0.81
1	0	0.96	0.99	0.82	0.98
0.5	0.5	1.00	1.00	0.83	0.99

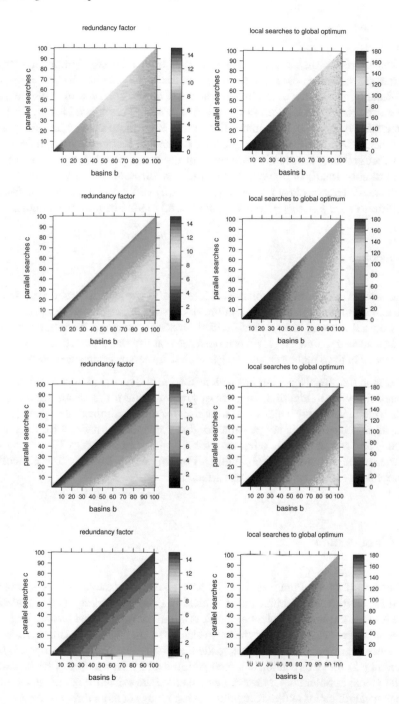

Fig. 3.4 Left: measured redundancy factors (t_3/b), averaged over $1,000$ repeats; basin size difference at most $1 : 10$. Right: local searches needed to locate the global optimum (t_2), derived from the same simulation. The rows from top to bottom: $p_{BI}, p_{BR} = (0,0), (0.5,0), (1,0), (0.5,0.5)$.

Observations

The most striking difference is that t_2 and t_3 both increase significantly, but at different rates: while t_2 rises about 30%, t_3 approximately doubles for $p_{BI} = p_{BR} = 0$. Additionally, we have a much higher variance, so the contours of different value ranges are more blurred. Both observations are more pronounced for higher basin numbers b. For $p_{BI} > 0$, the contour lines seem to be more skewed, meaning larger differences already for smaller values of c. For the last configuration including basin recognition, the overall difference in the corresponding equal basin result is remarkably small. In Table 3.3, we find few differences; only the fraction of significant differences in t_2 for $p_{BI}, p_{BR} = (0.5, 0)$ gets a bit larger, and the fraction of differences for $p_{BI}, p_{BR} = (0.5, 0.5)$ compared to 90% of t_2 is much higher.

Discussion

The larger variances can be explained at least partly with the different basin sizes of the basin that holds the global optimum; due to our experimental setup, this basin may take all allowed sizes, which will dramatically change the number of local searches needed to find it (t_2). We can presume that variances as well as the observed differences in the equal basin setting get higher for larger basin size contrasts.

Concerning the effectiveness of basin identifcation and recognition, we find that both, but especially basin identification, are of greater value if the basins are unequally sized, which is probably a realistic assumption for most optimization problems. From the available results, it is not possible to give concrete estimations for the rates at which t_2 and t_3 will increase other than for a basin size contrast of $10 : 1$. However, it is reasonable to expect that obtaining all optima will get much more difficult for higher values and that t_2 is much less affected.

3.4 Conclusions

Previous studies (e.g., Preuss et al. [177]) have shown that canonical EAs are not well suited for multimodal optimization. Are niching EAs? According to our simulations, there is some exploitable potential, but it is small for equal sized basins. It appears that chances get better the larger the basin size differences are. This situation changes a bit when many good solutions are needed, as the redundancy factor is much more affected by basin identification and recognition than by the number of local searches for the global optimum (t_2). However, we assumed the existence of efficient methods for basin identification and/or recognition, which may get more difficult to establish if search space dimensions and number of basins rise. Whether and for what problems such a technique can be fast enough to enable outperforming an ILS (with random restarts) remains to be seen.

Nevertheless, niching methods shall also be helpful in a different context. Namely, when insights about the treated optimization problem are to be gathered, of which the estimated number of optima as well as the basin size differences are important ones. Furthermore, it shall also be possible to successfully apply basin identification for multi-level (funnel) basin structures, by feeding it not arbitrary samples, but with the search space positions and objective function values of the optima obtained from local searches.

Chapter 4
Basin Identification by Means of Nearest-Better Clustering

Here we first collect the most important objectives for a basin identification (and thereby clustering) algorithm in the optimization context and then propose a technique for detecting clusters in populations of search points that correspond to basins of attraction. We present this method early and defer literature review and comparison, as it builds the basis for several measurements and algorithms that will be provided in later chapters.

4.1 Objectives

What properties are decisive for a clustering method that is suitable for online use as a basin identification method within optimization algorithms?

Since basin identification basically consists of detecting where the basins of attraction are located so that local optimization processes can be started, it appears obvious to attempt to gather this information without using too much additional effort. At the start of a population-based optimization algorithm, we usually have a more or less randomly initialized set of evaluated search points that may be employed for this purpose. Search points that become available later are dependent on the behavior of the optimization algorithm itself.

In the following, we presume that we can freely choose size and location of an initial sample of search points, and we apply some clustering technique to this sample. More concretely, a search point-to-cluster mapping is needed that resembles the search point-to-basin of attraction assignment implicitly given by the considered problem. If possible, this mapping should be bijective: only one cluster per basin of attraction and only one basin of attraction covered by a cluster. If the number of basins is small, we strive for a complete mapping, so that all of them are covered. In case the fitness landscape topology is known beforehand, which is usually the case for common benchmark problems, we may assess the quality of a clustering by means of this criterion. Generally, we strive for a clustering method that integrates with the optimization algorithm applied; cluster information gathered shall be used online, in contrast with approaches such as COGA [165], where the task is to provide interactive decision support to the user.

At this point, we may profit from insight gained in other branches of computer science dealing with clustering techniques. One particularly interesting field in this context is data mining. In data mining, cluster analysis often needs to cope with high-dimensional data, mixed types (e.g., real-valued, binary, nominal, or ratio variables), and very little knowledge concerning the appropriate number of clusters to find. The similarities in requirements when compared to basin identification within an optimization algorithm are striking.

From the viewpoint of a data mining professional, Berkhin names several important properties of a clustering algorithm [29]. Some of these are important here, too:

- Type of attributes an algorithm can handle.
- Ability to work with high-dimensional data.
- Ability to find clusters of irregular shape.
- Independence of [a priori knowledge and] user-defined parameters.

The first three correspond to difficulties often encountered during optimization: mixed type variables, high problem dimensionality, and non-spherical shape of basins of attraction. Thus they are specifications of the primary goal stated above. The fourth relates to the effort needed to obtain good results. It may happen that some problem-dependent parameter tuning is necessary. If so, this would downgrade the usefulness of a method.

Additionally, we explictly desire a *simple* clustering scheme; it shall be added as an auxiliary technique to an optimization algorithm and not in turn require some expensive tuning to the problem itself or expensive computation. Simplicity will also have benefits if we try to understand the interaction between the two.

In contrast with the data mining context, scalability of clustering algorithms used inside an EA is of minor significance: The number of data items is on the order of starting population sizes and most likely below $1,000$. Therefore, we give priority to quality and only demand a time complexity of at most $O(n^2)$. Lower orders and—of importance similar to that for bounded n—low constant factors are desirable, of course, but the relaxed time bound seems justifiable for real-world applications where a single fitness value computation is often much more costly than the rest of the optimization algorithm.

4.2 The Basic Nearest-Better Clustering Algorithm

What is the concept behind the nearest-better clustering technique?

We now present a clustering technique that is specifically tailored to sets of evaluated search points within an optimization process (coordinates plus objective function value) on a conceptual level; the details and concrete methods used are deferred to Sect. 4.3 to enable us to focus on the most important mechanisms first.

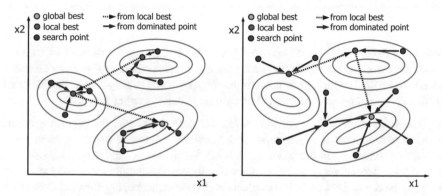

Fig. 4.1 Nearest-better clustering example; basins of attraction are indicated by contour lines. Search points connect to their nearest better neighbor. If they concentrate on different basins of attraction (left-hand side), the longest connections are the ones between local optima (dashed lines). For random search point distributions, distance differences can shrink, so that distinction of clusters becomes more difficult (right-hand side).

The *nearest-better clustering* algorithm (NBC) was first presented in [177] and is outlined in high-level pseudocode in Algorithm 2. It rests upon the assumption that the best yet sampled search points are approximations of different local optima and thus located in different basins of attraction. These should be much further away from each other than all search points are from their nearest better neighbor on average. Thus, the NBC algorithm first creates a spanning tree from the current population (see Figure 4.1, left hand side) by connecting each search point to the nearest one that is genuinely better. The longest edges are then cut, each time splitting a connected component into two. We identify these by means of a heuristic rule that uses a parameter ϕ. This parameter is multiplied by the mean length of all edges, and edges longer than the product are cut.

The first experimentation with varying ϕ led to the conjecture that it is much less sensitive to changes than a niche radius parameter would be (values around 2 seem to work well); however, this claim must be investigated further in the following.

There is one notable exception where executing lines 2 to 4 leads not to a spanning tree, but to a spanning forest. This is the case when there are multiple best points, so that the number of edges created within the **forall** loop is less than $n - 1$ for n search

---

Algorithm 2: Nearest-better clustering (NBC)

---
1 compute mutual distances between all search points;
2 create an empty graph with num(*search points*) nodes;
   `// make spanning tree:`
3 **forall the** *search points* **do**
4 $\quad\lfloor$ find nearest search point that is better; create edge to it;

   `// cut spanning tree into clusters:`
5 delete edges of length $> \phi \cdot$ mean(*lengths of all edges*) ;
   `// find clusters:`
6 find connected components;

---

points (by definition, there are no better points the best points could be connected to). However, the effect on NBC shall be negligible in most cases. In line 5 we consider all edges, be they connected or not, and having fewer edges only means that we start cutting from a number of connected components that is greater than 1.

An advantage of the NBC algorithm, at least as far as parameter tuning is concerned, is that it uses neither problem-dependent "*optimal*" distances nor a preset number of required clusters. On the other hand, it obviously needs a certain distribution of search points: Randomly scattered points usually contain cumulations, and these may wrongly be recognized as different clusters. In consequence, appropriate clustering of a completely randomly initialized search point population will be a much more difficult task for the NBC (Figure 4.1, right-hand side) than clustering an evolved population, or an evenly distributed start population. However, when it is applied at the right time, we can hope to detect at least some of the basins of attraction covered by a set of search points before the optimization algorithm concentrates on the most promising one.

The NBC algorithm is designed to strictly adhere to the objectives named in Sect. 4.1. It is therefore well suited to support population-based optimization algorithms because these deal with non-uniform data: a position vector and an assigned fitness value for each search point. The latter imposes a "*natural*" ordering on the items considered, which is exploited in the nearest-better step of the algorithm. In the general case, when such distinction is unavailable because there is no single most important attribute, NBC will not work at all. As a consequence, comparisons with well-known general clustering algorithms are possible only on test cases from the optimization field.

In the context of cluster analysis methods (see [78] for a recent overview), the NBC is a hierarchical clustering method. It is clearly based on connectivity, with the heuristic rule in line 5 deciding at which point to cut. Increasing the parameter ϕ would lead to less clusters, decreasing it would result in more.

4.3 Method Choice

Which concrete methods substitute the conceptual-level placeholders in the algorithm? What are their limitations?

Whereas the previous section introduced the nearest-better clustering algorithm on a conceptual level, we now have to fill the gaps to actually make it work. The steps of Algorithm 2 that need refinement are: distance computation, line 1, and connected components identification, line 5.

4.3.1 Distance Computation

In cluster analysis, it is more common to use similarity measures instead of distance functions; however, transforming one into the other is usually straightforward [95]. In evolutionary computation, the latter concept is widely used, e.g., to measure diversity within or overlap between different populations [244]. Additionaly, its spatiality connotation simplifies our thinking of such things as basins of attraction to an intuitive way, which is why we stay with this type of measure here.

For defining distance metrics suitable for identifying clusters within the search space of an optimization problem, at least three tasks need to be fulfilled:

- Choose an appropriate base metric for each uniform search space type (real-valued, ordinal discrete, nominal discrete, binary).
- Normalize variables to balance different ranges or local optima frequencies.
- Aggregate metrics for different variable types in case of mixed-integer or other multiple component search spaces.

In the following, we will first consider the related literature on these topics before recommending a few measures for further testing. We cannot hope to find a one-fits-all metric but strive for a robust solution that often works. It shall be emphasized that the NBC algorithm only needs a distance matrix and does not work on the original spatial location (coordinates), so we only need a distance metric to make it run even for very exotic data types such as those used in structure optimization. The metrics discussed in the following are recommendations for the more usual cases.

4.3.1.1 Base Metrics

Wineberg [244] discusses metrics based on the L_p-norm (Equation 4.1) for real-valued search spaces. He favors Euclidean distances because evolutionary algorithms mostly utilize "*Gaussian*" mutation operators in such environments, and the standard deviation rests upon the L_2 (Euclidean) norm. For binary search spaces, he justifies continued use of the Hamming distance, which is the metric traditionally employed in GA literature. According to his work, a "*true*" operator-based distance would be more appropriate, but the Hamming distance serves reasonably well as a first-order approximation of this "*true*" distance.

$$d_p(x,y) = \left(\sum_{i=1}^{D} (x_i - y_i)^p \right)^{\frac{1}{p}} . \qquad (4.1)$$

Grabmeier et al. [95] suggest use of the Mahalanobis distance (Equation 4.2, where B is the estimated covariance matrix) in case of different variable ranges. In its simplest form, starting from a diagonal matrix and choosing the standard deviations as entries, this distance weights the variables by the inverse of their variances. In its original form, the complete estimated covariance matrix is used, thereby reflecting interdependencies of the variables. Although theoretically appealing, this requires to some effort in data preparation before the metric gets actually usable. Therefore, we consider the Mahalanobis distance only as a last resort if all else fails. However, an advantage of this metric is that normalization, at least according to the search space volume currently covered by a population, becomes unnecessary.

$$d_M(x,y) = \sqrt{(x-y)^t B^{-1}(x-y)} . \qquad (4.2)$$

Most of the clustering-based optimization approaches known to us focus on rather low-dimensional problems; increasing search space dimension seems to downgrade the performance of basin identification methods until they eventually become useless. Facing a similar problem in data mining, Beyer et al. [32] voice doubts about the applicability of standard (L_p norm) distance metrics in detecting nearest neighbors in the high-dimensional case. They show that for a wide variety of data distributions, search for the nearest neighbor gets increasingly meaningless from about $D \approx 15$ (D means dimension) because the contrast, the ratio between largest and smallest existing distance, approaches 1.

Aggarwal et al. [4] report that for nearest neighbor detection, the L_1 norm often yields better results than the L_2 norm. Encouraged by this finding, they extend the L_k norm to fractional distance (semi-)metrics ($p < 1$), thereby enlarging the contrast for high-dimensional data. Our preliminary tests, however, have shown that using "*exotic*" measures leads to advantages in terms of accuracy only in very rare cases, which is why we disregard this alternative for now.

4.3.1.2 Variable Normalization

If the variables of an optimization problem have very different ranges, normalization may simplify clustering. In effect, variables are weighted equally with respect to the distance measure employed. However, normalization depends on given limits. For the unbounded case, or when the "*interesting*" (most searched) area of the search space is much smaller than these limits, we may be better off using the extension of the current search point population as the basis for normalization.

Nevertheless, most test problems, and many applications, set equal limits to the single variables, which makes normalization unnecessary. Thus, we shall apply it only if one of the two conditions described above are met.

4.3.1.3 Aggregation of Mixed Type Data

If the optimization variables are of different types, e.g., real-valued, discrete and Boolean, with the latter two partitioned into simple and structure-changing (indicating) variables, an aggregation method must be defined to ensure a meaningful distance matrix. Clearly, measuring distances between variables of different types is difficult and may not give results suitable for our clustering algorithm.

Grabmeier et al. [95] suggest computing the distances in type groups and adding them together with user-given weights. We shall adopt this approach because it is pragmatic and set the weights accordingly to the number of items in each group, resulting in approximately equally weighted variables. It may be tempting to rate indicating variables higher than others. However, in doing so we may miss clusters of search points that have a lot of simple variables in common and differ only in a structure-changing variable. After all, clustering by use of indicating variables is simple because they are known as important. No special clustering algorithm is needed to do a separation here.

Nevertheless, we shall emphasize that NBC can be applied as soon as a distance/similarity measure is available, even if the problem dealt with has non-real variables only.

4.3.2 Mean Value Detection

Line 5 of the suggested Algorithm 2 requires computing the average edge length of all edges. Out of the many ways to determine a mean value, at least two appear suitable here, the arithmetic mean, and the median. If the length distribution is almost uniform, resulting differences shall be small. For other distributions, much depends on location and size of the "*gap*" between the group of basin interconnections and local connections. The clustering algorithm must strive to estimate this gap from the edge length distribution because the edges beyond it must be deleted to split the population based on the underlying basins of attraction.

In the experimental comparisons we have made to detect if it is advantageous to choose mean or median, it turned out that the differences are usually rather small. For reasons of simplicity, we employ the median in the following, but the arithmetic mean would also do.

In the context of determining methods for a specialized clustering algorithm, one could think of a third alternative for estimating the gap. Assuming two distinct groups of edges, a simple standard clustering technique like 2-means may perform well. However, such a technique would fail when applied to a unimodal optimization problem because it is determined to always find two clusters, even when there are none in the utilized data. We therefore disregard it in the following.

4.3.3 Connected Components Identification

The components remaining connected after deleting "long" edges are the clusters we are interested in. To identify them in an abstract graph, often depth-first search (DFS) or breadth-first search (BFS) algorithms are employed. As the number of edges in our case is limited to at most $n-1$, both may be applied and shall lead to similar results in terms of time complexity, namely $O(n)$.

4.4 Correction for Large Sample Sizes and Small Dimensions

Do we need a special treatment to cope with the randomness of initial populations?

Preliminary experiments on smaller (two to five) dimensional problems have indicated that NBC often overestimates the real number of basins when the sample population is not very structured yet. This finding is backed by the experimental investigation carried out in [220]. Within a large number of randomly placed search points, we can regularly observe more crowded regions that can lead to wrongly determined basins. Imagine the situation where the initial set contains two heaps on opposite sides of a basin. It may happen that NBC does not detect that the two sets belong together. However, this mistake does not happen very frequently and we like to emphasize that in contrast to other clustering methods, NBC still succeeds here with high probability, due to its taking the objective values of the search points into account.

4.4.1 Nearest Neighbor Distances Under Complete Spatial Randomness

Overestimating the real number of basins means that a number of clusters are generated within one basin of attraction. This is especially troublesome for unimodal problems, which cannot safely be recognized as such. The effect seemingly gets stronger for larger point sets, most likely due to the growing expected maximum *nearest neighbor distance* (NND) for larger samples. A correction factor is therefore in order to take this statistical effect into account. To avoid too many false positives (basin detections where there are none), we need to determine the ratio of the expected largest nearest neighbor distance to the mean nearest neighbor distance under the hypothesis of *complete spatial randomness* (CSR) in unit space,[1] for a given sample size n and number of dimensions D. We are then able to modify the NBC in a way that connections shorter than or equal to this ratio, multiplied by the concrete sample's mean nearest (better) neighbor distance, are not cut, as they may just reflect the influence of randomness but have nothing to do with the actual objective function. This implies the assumption that the ratio is not much different for nearest *better* neighbor distances than it is for nearest neighbor distances. In other words, we make the NBC less sensitive to account for the randomness in the positioning of a population in the search space. Of course, this correction is not necessary for populations with a fixed spatial structure such as grid samples. However, grid samples are not very flexible in size, so for higher dimensions, only few distinct layouts are available. In order to have more possibilities for configuring an initial sample and to make it easier to integrate the concept into existing algorithms, we disregard grid samples for now.

Unfortunately, the desired largest nearest neighbor distance, also termed *connectivity distance* by Appel and Russo [11], is unavailable for Euclidean spaces and was derived only for the L_∞ distance norm. Furthermore, Diggle [63] states that the theoretical distribution of the nearest neighbor distance under CSR is not expressible in closed form due to complicated edge effects. We thus aim for a reasonable approximation according to his guidelines for the two-dimensional case.

4.4.2 Obtaining an Approximate Nearest Neighbor Distance Distribution Function

CSR is the basic property tested against in spatial (usually two or three-dimensional) statistics (see Diggle [63]). If the CSR hypothesis cannot be rejected, a sample is undistinguishable from a uniform distribution sample and searching it for any structure is doomed to fail. Needless to say, several test procedures have been es-

---

[1] hypercube spanned by vectors of unit length in each dimension

tablished to determine exactly that. It is most desirable to attain an (approximate) distribution function $G(r)$ for the nearest neigbhor distance r, as many test statistics and other statistical properties can be derived from it. Generalizing from Dimple's two-dimensional formulas, we obtain a first approximate distribution function ($G_{1V}(n,D,r)$, Eq. 4.5) with $S(D,r)$ being the volume of a hypersphere in D dimensions with radius r, and V meaning the total search space volume. In (4.4), $\Gamma(x)$ means the gamma function that is an extension of the factorial function and defined over all real (and most complex) numbers.

$$G_1(n,D,r) := 1 - \left(1 - \frac{S(D,r)}{V}\right)^{n-1} \overset{V=1}{\Rightarrow} G_{1V}(n,D,r) := 1 - (1 - S(D,r))^{n-1}.$$
(4.3)

$$S(D,r) := \frac{2\pi^{\frac{D}{2}} r^D}{D\Gamma\left(\frac{d}{2}\right)}.$$
(4.4)

$$G_{1V}(n,D,r) = 1 - \left(1 - \frac{2\pi^{\frac{D}{2}} r^D}{D\Gamma\left(\frac{D}{2}\right)}\right)^{n-1}.$$
(4.5)

This distribution function (Eq. 4.5) is numerically difficult to treat, as the huge exponent enlarges any rounding error very rapidly. However, extending a simpler two-dimensional exponential-based approximation for large n to the D-dimensional case leads us to (Eq. 4.6), which is much easier to handle numerically. Experimental comparison shows that for $n \geq 20$, the two functions differ only by a small margin.

$$G_{2V}(n,D,r) := 1 - exp\left(\frac{-n2\pi^{\frac{D}{2}} r^D}{D\Gamma\left(\frac{D}{2}\right)}\right).$$
(4.6)

From (Eq. 4.5) or (Eq. 4.6), the expected mean nearest neighbor distance for any n and D may be computed by integrating over the first derivatives (probability density functions, here denoted by G'), multiplied by r (Eq. 4.7). Note that the expectation may not exist if the integral is not absolutely convergent (the integral with r replaced by $|r|$ does not converge).

$$E(r) = \int_{-\infty}^{\infty} rG'(r)dr.$$
(4.7)

While it is still possible to give the derivatives in closed form, this cannot be done for the integrals. The same holds for the expectation of the largest NND that can be derived from the first-order statistics of the two equations.[2] Thus, this path to an

---

[2] by applying $G_{(1:n)}(r) = (G(r))^n$

approximation of the desired quotient appears disproportionately hard. An alternative approach could start from the connectivity distance (here denoted by c) formula of Appel and Russo [11] for $D \geq 2$:

$$\lim_{n \to \infty} \left(c^D \frac{n}{\log(n)} \right) = \frac{1}{2D} .$$ (4.8)

They have shown that for $n \to \infty$, the connectivity distance approximates the largest nearest neighbor distance, so that for large n,

$$E_{maxNND}(n, D) \approx c \approx \sqrt[D]{\frac{\log(n)}{2Dn}}$$ (4.9)

shall be on the order of the expected largest NND. However, the formulas have been derived for the L_∞ distance norm and it is not clear how large n has to be to get a reasonable approximation. As this approach only deals with the largest NND in a point set, we additionally need an approximation for the average NND to obtain a correction factor that "*filters out*" the influence of spatial randomness.

A rough approximation for the mean NND may be obtained by the following simple reasoning. Let the total volume of a D-dimensional box constrained space be 1, and let each of the n points contained in it have equally sized hypercubes around it that fill the whole volume. Then each of these boxes has volume $\frac{1}{n}$ and the edge size of the hypercubes is given in (Eq.4.10), which also resembles the minimal distances between two points as "*box centers*" if they are directly adjacent.

$$E_{meanNND}(n, D) \approx \sqrt[D]{\frac{1}{n}} .$$ (4.10)

This approximation is not bad for large dimensions, but lacks accuracy if $D \leq 5$. In the following experiment, we first review the quality achieved via the first approach described above. Then we compose equations (4.9) and (4.10) and correct it via experimentally derived factors to obtain a precise yet simple approximation for the sought correction factor of $E_{maxNND}(n, D)$ divided by $E_{meanNND}(n, D)$.

4.4.2.1 Experiment 4.1: Determine Maximum to Average NND Approximation

Pre-experimental Planning

To get a first impression of the approximation quality achieved via the two distribution functions (Eq. 4.5) and (Eq. 4.6), we compare the obtained expected values for the average and maximum NND with each other and with the results of a Monte Carlo simulation in Table 4.1. The resulting maximum to average factors are depicted

Table 4.1 Expected average and maximum NND as computed by means of the approximations given in (Eq. 4.5) ($E(G_{1V})$ and $E(\max(G_{1V}))$) and (Eq. 4.6) ($E(G_{2V})$ and $E(\max(G_{2V}))$) and as determined via Monte Carlo simulation. Note that the approximations do not account for boundary effects. Labels n and D mean population size and number of dimensions.

n	D	$E(G_{1V})$	$E(\max(G_{1V}))$	$E(G_{2V})$	$E(\max(G_{2V}))$	$\varnothing$(MC)	max(MC)
10	2	0.160	0.317	0.158	0.299	0.183	0.389
10	3	0.260	0.400	0.257	0.404	0.310	0.527
10	5	0.419	0.547	0.416	0.554	0.514	0.752
10	10	0.691	0.795	0.688	0.799	0.913	1.158
20	2	0.113	0.230	0.122	0.236	0.125	0.287
20	3	0.205	0.341	0.205	0.346	0.235	0.444
20	5	0.365	0.536	0.362	0.504	0.440	0.690
20	10	0.644	0.760	0.642	0.763	0.832	1.100
50	2	0.071	0.166	0.071	0.168	0.075	0.208
50	3	0.151	0.273	0.151	0.276	0.169	0.347
50	5	0.302	0.438	0.301	0.441	0.351	0.595
50	10	0.586	0.712	0.586	0.713	0.733	1.010
100	2	0.050	0.127	0.050	0.128	0.052	0.151
100	3	0.119	0.229	0.119	0.230	0.129	0.287
100	5	0.262	0.394	0.262	0.395	0.299	0.529
100	10	0.547	0.675	0.547	0.676	0.670	0.943

in Fig. 4.2. We find that the two approximations do not differ by much and that they consistently underestimate the measured values. The relative error seems to be largely independent of n and D and can most likely be attributed to boundary effects. Thus the quality obtained from the approximations does not render them unusable; however, their computational cost does, as they require two numerical integrations each (which take some seconds for the computer algebra system Maxima[3] to compute). As the approximations are needed for fast correction of measured nearest better neighbor distances in the NBC algorithm, the speed requirement outweighs the quality requirement by far. In consequence, we have to look for a simpler alternative.

Task

Starting from Equations (4.9) and (4.10), we want to derive a simple approximation for the maximum to average NND factor that gives less than about 10% average relative error for a meaningful range of n and D.

---

[3] Maxima is available under GPL license at http://maxima.sourceforge.net/

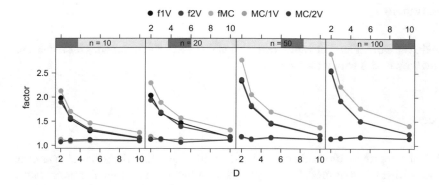

Fig. 4.2 Maximum to average NND factors for the two approximation approaches (f1V is computed via (Eq. 4.5), f2V via (Eq. 4.6)) and as obtained from Monte Carlo simulation (fMC). Note that the approximations ignore boundary effects. MC/1V and MC/2V denote the quotients of simulated and approximated factors. The labels n and D stand for population size and dimensionality.

Setup

We set the range of n to $10 : 300$, tested in steps of 10, and the range of D to $1 : 30$, every value tested. At first, we generate the *"real"* maximum to average NND factors by Monte Carlo simulation with 100 repeats for each of the $n \times D$ combinations (average and maximum values are simulated separately and then divided). This resembles an extended sample of the one given in Table 4.1 (MC columns). We then apply the approximation of (Eq. 4.11) to every entry of the $n \times D$ table and compute the average relative error after equation (Eq. 4.12), where $\frac{\max(MC(n,D))}{\text{mean}(MC(n,D))}$ stands for the Monte Carlo-determined (maximum to average NND) correction factor.

$$cfNND_1(n,D) = \frac{E_{maxNND}(n,D)}{E_{meanNND}(n,D)} \approx \frac{\sqrt[D]{\frac{\log(n)}{2Dn}}}{\sqrt[D]{\frac{1}{n}}} = \sqrt[D]{\frac{\log(n)}{2D}} \,. \tag{4.11}$$

$$\text{err}_{cfNND_1} = \text{mean} \left| \frac{cfNND_1(n,D) - \frac{\max(MC(n,D))}{\text{mean}(MC(n,D))}}{\frac{\max(MC(n,D))}{\text{mean}(MC(n,D))}} \right| \,. \tag{4.12}$$

Result

As relative approximation error of $cfNND_1(n,D)$ we obtain 35.49%.

Observations

Looking at the single error values reveals that these rapidly get smaller for increasing D and slowly rise with increasing n.

Discussion

While the approximation obviously points in the right direction, the relative error is much larger than desired. The single errors show a need for an additional factor that somehow depends on D but grows slower than linear. We therefore refine our formula and add some degrees of freedom, thus obtaining an optimization problem for the best approximation.

Refined Setup

In order to achieve better precision, we introduce a factor of $a\sqrt[b]{D}$ that contains two unknowns, a and b. To obtain a good setting for these two variables, we perform a grid test in the ranges $a = 1 : 5$ and $b = 2 : 6$ with 20 steps per unit, resulting in 81×81 test points. For each test point, we measure the average relative error against the 30×30 n and D configurations as described above. From the best test point, the BFGS implementation of Nash [157] in the statistical software R is run as a local optimization tool.

Result

With an average relative error of 2.924%, $a = 2.95$ and $b = 4$ is the best of all tried grid configurations. Successive BFGS application leads to minor improvements only, yielding 2.923% error.

Observations

The obtained best configuration is very robust against slight parameter changes, e.g., 196 other a and b combinations (out of 6,561) return less than 5% error; 512 more stay below 10%.

Discussion

We finally end up with a relatively simple formula for the correction factor as given in (Eq. 4.13). Attempting even better accuracy seems inappropriate as the Monte Carlo simulated values we try to approximate are themselves not exact. Additionally, the obtained accuracy shall suffice as we only aim at ruling out largely overestimating the number of clusters. Note that the correction factor only applies if CSR holds, that is, at the very beginning of a population-based optimization process, which is started with a random sample.

$$cfNND_2(n,D) = 2.95 \sqrt[4]{D} \sqrt[D]{\frac{\log(n)}{2D}} . \tag{4.13}$$

4.4.3 When to Apply the Correction

Using this correction factor within the NBC makes it less prone to detecting too many clusters by simply decreasing its sensitivity. It applies to line 5 of Algorithm 2; instead of cutting edges of length $> \phi \cdot$ mean($lengths\ of\ all\ edges$), we cut edges of length $> cfNND_2(n,d) \cdot \phi \cdot$ mean($lengths\ of\ all\ edges$). This artificially degrades the possibility of finding many clusters and shall only be applied to random samples (CSR holds).

4.5 Nearest Better Clustering Extended With a Second Rule

What other properties of the established tree can be used to improve the clustering?

As previously described, the basic rule (in the following also called rule 1) of NBC works by connecting every search point in the population to the nearest one that is better and cutting the connections that are longer than ϕ times the average connection. The remaining connections determine the clusters by computing the weakly connected components. Rule 1 works very well for reasonably large populations in few dimensions, but increasingly fails if the number of dimensions increases, as the analysis of the results of [170] has shown.

In [171], we have therefore added a second additional cutting rule: For each search point that has at least three incoming connections (it is the nearest better point for at least three others), we divide the length of its own nearest-better connection by the median of its incoming connections. If this is larger than a precomputed correction factor b, the outgoing connection is cut (and we have one additional cluster). Both rules are applied in parallel, that is, the edges to cut due to rule 2 must be computed

---

Algorithm 3: Nearest-better clustering (NBC) with rule 2

---

1 compute all search points mutual distances;
2 create an empty graph with num(*search points*) nodes;
   `// make spanning tree:`
3 **forall the** *search points* **do**
4   ⌊ find nearest search point that is better; create edge to it;

   `// cut spanning tree into clusters:`
5 **RULE1:** delete edges of length $> \phi \cdot$ mean(*lengths of all edges*);
6 **RULE2: forall the** *search points with at least 3 incoming and 1 outgoing edge* **do**
7   **if** *length(outgoing edge)/median(length(incoming edges))* $> b$ **then**
8   ⌊ cut outgoing edge;

   `// find clusters:`
9 find connected components;

---

before actually cutting due to rule 1. Edges cannot be cut more than once, so if both rules apply, this is not specially treated. Algorithm 3 presents the updated NBC method containing both rules. As the basin identification capability of rule 1 in 2D seems to be sufficient, rule 2 is only applied if $D \geq 3$.

4.5.1 Deriving a Correction Factor For Rule 2

The motivation for rule 2 was that in sufficiently large samples (at least around $40 \times D$, where D is the number of dimensions), we often find points with several incoming connections whose outgoing edge is not cut with rule 1 because it is longer than all the incoming ones but not one of the longest of the whole sample. However, determination of the correction factor b for rule 2 is not trivial. Here, it is experimentally derived and presumed to depend on D and the sample size n. As we want to recognize only one cluster on unimodal problems and ideally two or more on multimodal problems, we employ two extreme test functions. These are a sphere with the optimum aligned to the center of the search space (4.15) and a deceptive function (4.14) with 2^D optima, located in the corners of the search space (restricted to the hypercube $[0, 1]^D$ in both cases):

$$dec(\mathbf{x}) = \sum_{i=1}^{D} 1 - 2 * abs(0.5 - x_i) \,, \qquad (4.14)$$

$$sphere(\mathbf{x}) = \sum_{i=1}^{D} (x_i - 0.5)^2 \,. \qquad (4.15)$$

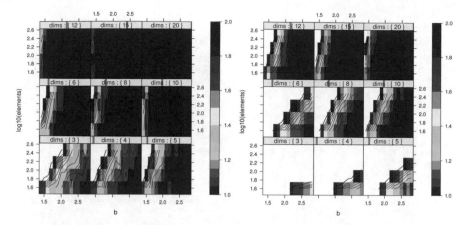

Fig. 4.3 Number of clusters found for 40 ($\approx 10^{1.6}$) up to 300 ($\approx 10^{2.5}$) elements (n) in dimensions three to 20, by applying different values for correction factor b. Left: sphere (we must not obtain more than one cluster here); right: deceptive test function with 2^D basins. b has been chosen to result in at most 1.1 clusters on average on the sphere. The gap with the corresponding value for the deceptive function shows that still at least two clusters can be found reliably here up to around 20D. White areas mean values > 2.

In the following, we experimentally develop how to set the factor b. Performance comparisons using rule 1 alone on the deceptive function can be found in [171]; we only provide a visual example here in Figure 4.4 (left).

4.5.1.1 Experiment 4.2: Determine Correction Factor b for Rule 2

Pre-experimental Planning

Figures 4.3 and 4.4 (right) show the number of detected clusters for different settings of b over different dimensions (NBC with only rule 2 enabled, averaged from 100 repeats, $\phi = 2$). We find that it is possible to set b to a value such that at most 1.1 clusters are detected for the sphere problem, but a larger number are for the deceptive and the Rastrigin function. However, the corridor is shrinking for higher dimensions.

Task

We want to find a setting for b that recognizes exactly 1.1 clusters on average on the sphere function (assuming that this is the maximum error we are willing to accept).

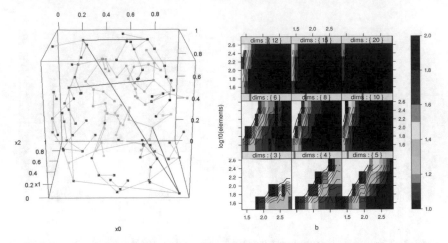

Fig. 4.4 Left: example clustering of a uniform randomly spread 120 point sample on the 3D deceptive test function; points are colored per detected cluster, gray lines are the nearest better edges, blue lines represent the edges cut due to rule 1, red lines the ones cut due to rule 2. The eight global optima are located in the corners of the cube. Right: the same setup as for Figure 4.3, but for the Rastrigin function.

Setup

For uniform randomly sampled populations of 40 and 300 elements, we apply a simple (non-deterministic) interval search method for some dimensions between *three* and 30. Starting from an educated guess on the basis of Figure 4.3, left, we try to decrease the interval by setting a middle point and repeating the measurement until we reach an upper limit or can detect if the required result of 1.1 clusters is located in the lower or upper half of the interval. The minimal number of repeats is set to 1000, the maximal number to 5000.

From Figure 4.3, we already know that for logarithmic scaling in the number of elements, the sought b values for obtaining a cluster number of 1.1 largely have a linear shape. Therefore, we obtain simple regression equations for 40 and 300 elements (separately) via a linear regression over the number of dimensions (method `lm` from the statistical software R) in a second step. The b values for other element numbers are then estimated via linear interpolation over $\log 10(S)$.

Result

The resulting estimations for 40 and 300 elements are shown in Figure 4.5 (left) as dot sets (the upper set of dots corresponds to 300 points). Linear regression results in Equations 4.16 and 4.17, which are used to draw the upper and lower line,

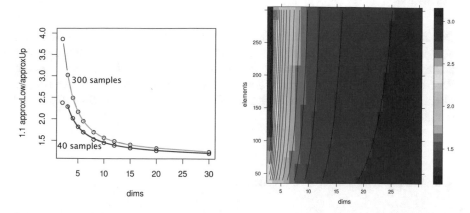

Fig. 4.5 Left: estimations for b in order to reach 1.1 clusters on the sphere function, for 300 and 40 elements. The orange curve covers the 300 element samples and resembles a linear regression over D, the blue curve is made similarly from the 40 element samples. Right: values for b as obtained from the final regression-based equation (4.18)

respectively. Employing these two equations for a linear interpolation over $\log 10(n)$, we obtain the more general Equation 4.18. Figure 4.5 (right) shows computed values for b as obtained from (4.18).

$$b(300,D) = 6.8/D + 0.7 + 0.023 * D + -4.1 * 10^{-4} * D^2 \,, \qquad (4.16)$$

$$b(40,D) = 3.6/D + 1.1 \,, \qquad (4.17)$$

$$b(n,D) = (-4.69 * 10^{-4} * D^2 + 0.0263 * D + 3.66/D - 0.457) * log10(n)$$
$$+7.51 * 10^{-4} * D^2 - 0.0421 * D - 2.26/D + 1.83 \,. \qquad (4.18)$$

Observations

Considering the three functions in Figures 4.3 and 4.4 (right) again, it seems that the deceptive function is relatively well suited for applying rule 2; for the Rastrigin function the differences to the sphere are already much smaller.

Discussion

Our correction factor approach for rule 2 is carefully designed to prevent detecting non-existent basins, especially on unimodal functions. However, by avoiding false positives, we may sacrifice good results on highly multimodal functions (as for the Rastrigin function). In any case, we find again that clustering and thereby basin identification is getting increasingly difficult in high dimensions; $D = 20$ already seems to be near the limit.

4.6 Measuring NBC Performance

How well does NBC perform with different numbers of available evaluated search points? How do we set the parameter ϕ and what initial distribution and correction do we use?

We have specified the basic algorithm and provided concrete methods for most of its variable parts. However, the question for the parameter ϕ of our method remains. Shall it be set to 2.0 by default, regardless of the problem and its dimensionality? Furthermore, we have provided two possible extensions of the basic NBC algorithm, namely the randomness correction for large, uniformly spread search point populations, and the rule 2 extension which at least for some cases (as shown in [171]) considerably improves the clustering quality. But under what conditions does it actually make sense to use these extensions?

We will approach these questions by means of experimentation, as the number of degrees of freedom would render any theory-based approach extremly difficult. As previously stated (Sect. 4.2), the nearest-better clustering algorithm is not applicable to standard clustering benchmarks; these fail to provide quality information for each data item, which is essential for NBC to work. Instead, test cases resemble realistic distributions of search points found within optimization processes. Therefore, we first introduce the test problems and performance measures we use for evaluating NBC. However, before that we provide a common model these parts may utilize.

4.6.1 Populations, Basins of Attraction, and Clusters

We postulate that a very simple model is sufficient for describing the interplay between clustering methods and optimization algorithms insofar as it enables assigning clusterings reasonable performance values. Using constructs from set theory, we can formulate relationships between the constituents of our model without specifying a concrete search space type. Nevertheless, we may assume we are dealing with multiple (few) dimensions of real numbers to support graphical intuition.

Our simple model consists of the following elements:

$X, \mathbf{x}$ The search space X is a set that contains all valid search points $\mathbf{x}$.

P_t Optimization algorithms and clustering methods work with successive populations of search points we denote by $P_t = \{\mathbf{x}_1, \ldots, \mathbf{x}_n\}$, where t is a time index and n the cardinality of P_t.

$\mathscr{C}, C_i$ A clustering $\mathscr{C}$ is a decomposition of a population P_t into a set of subsets C_i, so that $\mathscr{C} := \{C_1, \ldots, C_m\}$ and $P_t = \bigcup \mathscr{C}$. When the clustering is crisp, as opposed to fuzzy, the subsets are disjoint and the decomposition is a partition.

$\mathscr{B}, B_i$ We assume that the whole search space consists of a number of basins of attraction, denoted by sets B_i, and possibly a basin-free space in between. The latter is termed B_0 and need not be connected. The set system made up from all basins and B_0, denoted *basin system*, with $\mathscr{B} := \{B_0, \ldots, B_k\}$, then extends to the whole search space. That is, for every $\mathbf{x} \in X$ there exists $0 \le i \le |\mathscr{B}|$ so that $\mathbf{x} \in B_i$.

In addition, we define some functions:

$\mathrm{basins}(P_t)$ As we cannot hope to detect basins where there are no search points, we need a function for detecting the number of basins containing at least one search point from population P_t. The subsitute basin B_0 is deliberately excluded here:
$$\mathrm{basins}(P_t) := |\{B \in \mathscr{B} \setminus B_0 | \exists \mathbf{x} \in P_t : \mathbf{x} \in B\}| \,.$$

$\mathrm{basin}(\mathbf{x})$ Returns the basin $B \in \mathscr{B}$ a search point $\mathbf{x}$ is located in:
$$basin(\mathbf{x}) := \{B \in \mathscr{B} | \mathbf{x} \in B\} \,.$$

$\mathrm{cluster}(\mathbf{x})$ Gives the cluster $C \in \mathscr{C}$ of a clustering a search point belongs to:
$\mathrm{cluster}(\mathbf{x}) := \{C \in \mathscr{C} | \mathbf{x} \in C\}$. In case of a fuzzy clustering, this should be the cluster it is most associated with.

4.6.2 Test Problems

For testing in the real-valued domain, we employ a random multiple peaks problem generator, a slightly modified version of the one proposed in [173]. The generator is highly configurable and produces moderately multimodal, non-symmetric, and non-decomposable problems. These can be expected to scale worse than linear in terms of function evaluations in the problem dimension when tackled with an evolutionary algorithm (compare [195]). Note that except for the shape of the single peaks (polynomials instead of Gaussians), the generator is similar to the one suggested by

Yuan and Gallagher [249, 85]. We term this single-level generator *multiple peaks model* (MPM). In addition, we suggest a funnel-based extension of the MPM called *funnel MPM* (FMPM).

4.6.2.1 Multiple Peaks Model (MPM)

The basic working principle of the MPM generator is to randomly determine the locations of a number of peaks, compute the individual shapes of the peaks from a number of parameters, and determine where a sampled point hits the convex hull of the peaked landscape.

Table 4.2 Configuration of the real-valued multiple peaks problem generator. For the degrees, radii, and heights, ranges may be specified; concrete values are drawn randomly from uniform distributions over these ranges. Without loss of generality, the search space is limited to $[0,20]^D$. The resulting problem contains one global optimum (if all $h_b < 1$) and $k - 1$ local optima.

variable	parameter meaning	domain	default
D	problem dimension	$\mathbb{N}$	10
k	number of peaks	$\mathbb{N}$	20
p_{bi}	location of peak b in dimension i	$\mathbb{R}$	$p_{bi} \in [0,20]$
g_b	degree of polynomial (of peak b)	$\mathbb{R}_+$	$g_b \in [1,3]$
r_{bi}	peak b radius in dimension i	$\mathbb{R}_+$	$r_{bi} \in [10,20]$
h_b	height of peak b, relative to 1.0 (global best)	$\mathbb{R}_+, h_b < 1$	$h_b \in [0.9, 0.95]$
ROT_b	rotation matrix for peak b	$\mathbb{R}^{D*D}$	$-\frac{\pi}{4} < ROT_{bij} < \frac{\pi}{4}$

The obtained fitness landscapes resemble plateaus at value 1.0 with randomly scattered and shaped cones as depicted in Figure 4.6. The fitness domain is fixed to $[0,1]$, with 0.0 as global optimum, assuming minimization. Each peak b possesses radii r_{bi}, determining its maximum extension in every problem dimension $i \in \{1,\dots,D\}$ prior to rotation. As a modification from [173] in order to ease the parametrization in different dimensions, the radii are multiplied with the relative length of the space diagonal (of the unit hypercube) $\sqrt{D}$.

The peaks are shaped by a simple underlying polynomial of degree g_b and stretched to relative height h_b. We introduce variable interactions by means of rotation matrices ROT_b. Table 4.2 enumerates the parameters and provides default values employed for all experiments in this section, unless stated otherwise.

Fitness values of search points $\mathbf{x}$ are computed as follows: first, a corrected distance vector $\mathbf{c}_b$ for every peak $\mathbf{p}_b$ is determined by multiplying the coordinate vector with the rotation matrix ROT_b as denoted in Equation 4.19. We then calculate the relative Euclidean distance of the search point to each peak, compared to the maximum extent of the corresponding basin $\mathbf{r_b}$, and raise it to the g_bth power. As seen in Equation 4.20, the relative distance is 0.0 at the center of the peak and artificially limited to 1.0 at the border of the peak.

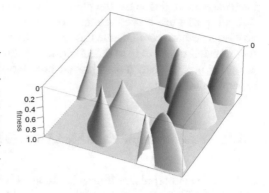

Fig. 4.6 Perspective view of a two-dimensional generated multiple peaks problem for minimization, seen from below to enhance visibility (best fitness is 0.0). Different peak shapes result from polynomials where degrees vary between 1 and 3. The uncovered search space is assigned fitness value 1.0. Axial symmetries are partly shifted from the coordinate axes due to variable interactions that lead to non-separability.

The fitness of $\mathbf{x}$ based on one peak results from multiplying the relative distance to a peak by the peak's height h_b. Out of all k values obtained for the different peaks, we take the minimal (best) one, correct it by adding $(1 - h_b)$, and return it as fitness function value $f(\mathbf{x})$ (Equation 4.20). The correction shifts the value of the border of the respective cone to 1.0, which is our upper limit.

$$\mathbf{c}_b = (\mathbf{x} - \mathbf{p}_b)^T ROT_b .$$
(4.19)

$$f(\mathbf{x}) = \min_{b=1}^{k} \left[\left(\min \left(\sqrt{\sum_{i=1}^{d} \frac{c_{bi}^2}{r_{bi}^2}}, 1 \right) \right)^{g_b} h_b + (1 - h_b) \right] .$$
(4.20)

The two main advantages of this approach are i) that it is straightforward to retrieve the attraction basin any search point lies in (it belongs to the peak that generates the minimum value in Equation 4.20, the one that finally becomes $f(\mathbf{x})$), and ii) availability of different, though related, problem instances. These advantages are shared with the generator suggested by Yuan and Gallagher, which uses Gaussian mixtures instead of polynomials and covariance instead of rotation matrices. We chose rotation because it appears slightly more intuitive and was also used by Salomon [195], but the same results may be produced by either technique.

Note that compared to most other test functions and problem generators, both approaches assume different variable interactions for each peak. Separate covariance/rotation matrices entail an important consequence for any search method that employs internal models of the problem treated: Information extracted from learning in one basin is not transferable to another. This property strongly necessitates basin identification. Otherwise, any form of globally applied adaptation may largely misguide the search process.

One can argue that many real-world problems probably exhibit similar variable interactions throughout the whole search space and thus the produced problems are overdesigned. In this sense, the suggested generator represents a domain of extremely difficult problems. Nevertheless, it is unsuitable for modeling highly multimodal problems because computation time grows linearly in the number of peaks. However, the intended use here is to detect clusters, a hopeless task if the number of basins is by far greater than the employed population sizes and no moderately modal superstructure exists.

4.6.2.2 Funnel Multiple Peaks Model (FMPM)

In order to also represent fitness landscapes with moderate global structure,[4] we extend the MPM generator by adding a second MPM with fewer, larger basins. Such funnel type problems have been created by Addis et al. [2], who build a generator with several levels, and Lunacek et al. [134], who assemble a multi-funnel problem from highly multimodal single problems.

Fitness computation of the FMPM is given by Equation 4.21. On the basis of an existing MPM_p, a second MPM_f is created for the funnels, and both are added using weight w_f. For all FMPM instances used here or in the following experiments, we set $w_f = 0.75$. The parameter set of MPM_f is otherwise very similar to the one of MPM_p, but the number of peaks (k from Table 4.2) should be much smaller than for MPM_p. There is another dependency of MPM_f from MPM_p that prevents creating new local optima unintentionally: all peaks of MPM_f must resemble MPM_p peaks (in location, not in shape or size). Figure 4.7 depicts two example instances with different funnel and peak numbers.

$$f(\mathbf{x}) = w_f MPM_f(\mathbf{x}) + (1 - w_f)MPM_p(\mathbf{x}) . \qquad (4.21)$$

4.6.3 Performance Measures

How do we measure the quality of a clustering in the context of optimization? A perfect solution would contain the same search point grouping that could be obtained if the basins of attraction were known. For the given test problem, the latter are well defined and performance criteria thus directly derivable from measured conformance. Three different types of discrepancy—or error—may occur between basins and decided clusters.

Type 1: Basins are not covered at all by any cluster; they remain undetected.

---

[4] This is unlikely to exist except for probabilistic anomalies when a low number of peaks is randomly scattered over the search space.

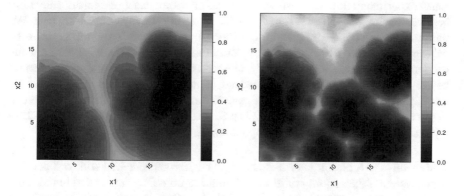

Fig. 4.7 Bird's eye view of a two-funnel 20 peaks 2D FMPM problem (left), and a five-funnel 100 peaks 2D FMPM problem. Lower values are better (minimization).

Type 2: Clusters cover more than one basin (cluster B in Figure 4.8).

Type 3: Many clusters reside in one basin (clusters B, C, and D in Figure 4.8).

Type 1 errors surely occur for basins lacking any search points, which is not the fault of the clustering method. In case only one search point is located in a basin, and if the clustering is complete (every point is assigned to a cluster), either it is detected as separate one element cluster or a type 2 error is produced because it is added to a cluster already containing points in other basins.

Whenever a cluster covers many basins (type 2 error), the clustering is too coarse-grained. Nevertheless, the cluster may be split appropriately if the clustering method is applied repeatedly in the course of optimization—there is no immediate loss of information. Type 3 errors are most undesired because optimization algorithms are led to making much effort for concurrent searches in the same basin. Our main interest is in the remaining non-errorneous clusters, because they properly cover one basin only.

Fig. 4.8 Schematic drawing illustrating possible errors during clustering; basins of attraction are indicated by contour lines. A cluster may spread over multiple basins (B) or several clusters may contain search points within the same basin (B, C, D). In this example, cluster A represents the most desired cluster type: All search points within a basin are assigned to the same cluster that does not spread onto other basins.

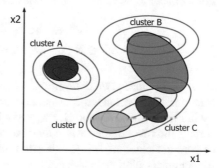

Early experimentation demonstrated that asking for an exact one-to-one matching between basins and clusters renders clustering an immensely demanding task. We therefore settled upon a pragmatic solution with slightly relaxed requirements. To prevent small groups of outliers from "*contaminating*" an otherwise pure cluster, we demand only that the majority of a cluster's constituents be located in the same basin. The choice of the majority instead of any other quantile is motivated by investigations concerning the effect of genetic drift on EA subpopulations that spread to more than one basin of attraction [177]: A small difference in size usually leads to rapid supersession of the smaller subpopulation.

Definition 4.1. A cluster C of a clustering $\mathscr{C}$ is termed *decided* with respect to a basin system if the majority of its constituent search points are located in the same basin of attraction. The corresponding basin is called its *main basin*. Clusters not meeting this condition are termed *mixed*.

We determine the main basin of a cluster with the following function and provide a predicate for detecting decidedness derived from it.

$$\text{mainBasin}(C \in \mathscr{C}) := \begin{cases} B & : \quad \exists B \in \mathscr{B} : |\{\mathbf{x} \in C : \text{basin}(\mathbf{x}) = B\}| > \frac{|C|}{2} \\ \emptyset & : \quad \text{else} \end{cases} \quad (4.22)$$

$$\text{isDecided}(C \in \mathscr{C}) = \begin{cases} \text{true} & : \quad \text{mainBasin}(C) \neq \emptyset \\ \text{false} & : \quad \text{else} \end{cases} \quad (4.23)$$

Definition 4.1 has been established to separate clusters that are subject to type 2 errors from ones that are not. However, it is necessary to also detect type 3 errors. That is, a decided cluster may be alone in a basin or accompanied by others.

Definition 4.2. A decided cluster C of a clustering $\mathscr{C}$ is termed *freestanding* with respect to a basin system if its main basin is different from the main basins of all other decided clusters in $\mathscr{C}$. Otherwise, it is called *surrounded*.

For a cluster to be freestanding, it has to fulfill the following predicate:

$$\text{isFreestanding}(C \in \mathscr{C}) := \text{isDecided}(C)$$
$$\wedge \, \nexists C' \in \mathscr{C} \setminus C : \text{mainBasin}(C') = \text{mainBasin}(C) . \quad (4.24)$$

By applying these two definitions, we may now partition a clustering $\mathscr{C}$ into three fractions, the freestanding, surrounded, and mixed clusters, represented in Figure 4.8 by items A, C and D, and B, respectively. Which of these clusters are useful and should thus be counted when evaluating a clustering? The intention of applying clustering techniques here is extraction of information concerning basins of attraction for later use within niching-based evolutionary algorithms. Undoubtedly, freestanding clusters provide such information, whereas surrounded ones are rather unwanted.

However, expecting a perfect match between clusters and basins may be unrealistic when we are confronted with a black-box optimization problem. Thus, if surrounded clusters occur in a clustering, they must be used to the best extent possible. By definition, joining all surrounded clusters with equal main basins would lead to a number of new freestanding clusters, one for each such basin. In consequence, it makes sense to count surrounded clusters with equal main basins as one useful cluster. Following from that, we can define the number of useful clusters of a clustering.

Definition 4.3. The *number of useful clusters* of a clustering $\mathscr{C}$ with respect to a basin system is defined as the number of basins covered by any decided cluster.

Note that this definition does not provide the useful clusters themselves; for measuring the quality of a clustering only their number is needed. It may be determined by means of the following equation:

$$\text{usefulClusters}(\mathscr{C}) := \left| \bigcup_{i=1}^{|\mathscr{D}|} \text{mainBasin}(C) \right|$$

$$\text{for } \mathscr{D} = \bigcup C \in \mathscr{C} : \text{isDecided}(C) \tag{4.25}$$

Parameterizing clustering methods mostly affects their sensitivity, resulting either in many clusters with few elements or few clusters with many; we have to find the right balance to achieve a suitable solution. Obtaining as many useful clusters as possible, but only these, may be viewed as two partly conflicting performance criteria, namely the fraction of useful clusters and the detected basin fraction, both to be maximized. This distinction has its roots in the heuristic nature of the method applied for estimating the true number of clusters. Based on the model laid down in Sect. 4.6.1, the two criteria may be expressed as follows:

$$\text{usefulClusterFraction}(\mathscr{C}) := \frac{\text{usefulClusters}(\mathscr{C})}{|\mathscr{C}|}, \tag{4.26}$$

$$\text{detectedBasinFraction}(\mathscr{C}) := \frac{\text{usefulClusters}(\mathscr{C})}{\text{basins}(\bigcup \mathscr{C})}. \tag{4.27}$$

In the following, the usefulClusterFraction and detectedBasinFraction functions are denoted by UCF and DBF, respectively. Note that the argument of the basins function in Equation 4.27, $\bigcup \mathscr{C}$, equals P_t and thus only depends on the distribution of the search point population, but not on the concrete clustering.

4.6.4 Variant Choice and Performance Assessment

We now tackle the questions asked at the beginning of Sect. 4.6, namely for the value of parameter ϕ and the usefulness of the proposed extensions (namely the correction for large sample sizes and the introduction of rule 2). At the same time, a first assessment of NBC performance shall be provided. The overall setting is the one of an envisioned initial population that must be clustered in order to concentrate the optimization process onto a number of most interesting regions, and to finally deliver several very good solutions representing the different basins of attraction. The latter may be achieved by different approaches, e.g., starting local searches in each cluster, or implementing mating restrictions or specific selection methods within a larger population of solutions.

There are at least two important degrees of freedom for measuring clustering performance, besides the chosen problem landscape, namely the number of dimensions and the population (sample) size. While it is in general not possible to attain detailed information about the optimized problem, the dimension is usually fixed and known. The sample size is a factor we can control when starting an optimization process. As we have already seen in Sect. 4.5, the dimensionality strongly influences the hardness of obtaining a suitable clustering and could thus also have an influence on the parametrization of the clustering method. Therefore, our general idea for the following experiment is to determine the most suitable variant per dimension. In order to perform this choice on a solid base, we employ several different problems. Nevertheless, the NBC is envisioned for moderately dimensional and moderately modal problems, and for the variant/parameter choice, we ignore all use cases that are far off from this problem type.

4.6.4.1 Experiment 4.3: Determine Dimension-Dependent NBC Variants

Pre-experimental Planning

Turning on and off three factors (correction, *Latin Hypercube Design* (LHD) or random initialization, and rule 2) results in eight different variants, and we obtain a first idea on how they influence the NBC performance in running all combinations on a five-funnel FMPM problem with 100 peaks. The task is to identify the funnels, not the single peaks. Figure 4.9 shows the UCF and DBF values for three representative factor combinations, namely:

a) with correction, random initialization, without rule 2,

b) without correction, random initialization, without rule 2, and

c) with correction, random initialization, with rule 2.

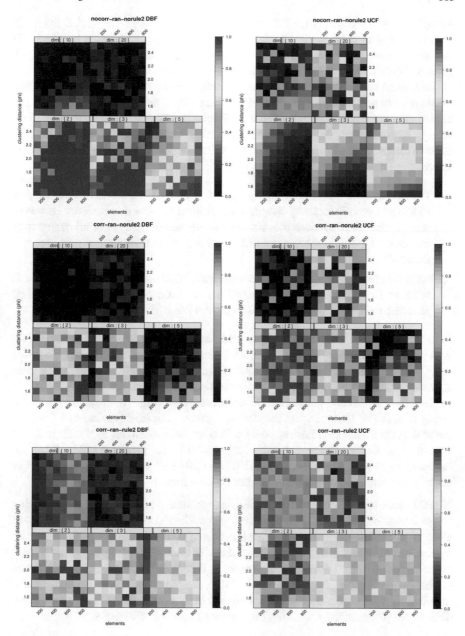

Fig. 4.9 Useful cluster fraction and discovered basin fraction for different NBC configurations on the FMPM problem generator with five funnels and 100 peaks, averaged from 100 repeats. Left column: detected basin fractions (DFB); right column: useful cluster fractions (UCF). Rows from top to bottom: no correction, random initialization, no rule 2; with correction, random initialization, no rule 2; correction, random initialization, with rule 2.

Sample sizes are varied from 100 to 800 and the ϕ parameter is varied from 1.5 to 2.5. Each data point is the averaged result of 100 repeats. Figures for the LHD initialization are not shown as they are largely similar to the ones for random initialization. The pictures for the configuration without correction, random initialization, and activated rule 2 resemble the ones for deactivated rule 2, except that the DBF values tend to be a bit higher and the UCF values a bit lower. The configurations a) to c) also clearly demonstrate that there is no single, dominating solution to the variant selection problem: choosing a variant rather means choosing a different point on a trade-off between maximizing DBF and maximizing UCF. However, we note some interesting effects:

1. A large sample size is not always good. The UCF values for configuration b) worsen with increasing sample size, at least for low dimensions. However, we may take into account that for very small sample sizes, not all basins may be covered, so the problem simplifies a bit (basins without any point are not counted).
2. Rule 2 activation, at least together with correction (configuration c)), seems to work especially well in high dimensions (ten and 20) although the achieved absolute fractions of detected basins (DBFs) are still low.

In order to increase the visibility of the trade-offs between achieved DBF and UCF values, we will change to a multi-criterial perspective in the following. However, as UCF and DBF values are both considered important, we have to aggregate them at some point, which is done by multiplication. Note that the obtained value is somewhat artificial and does not have a concrete meaning. We only know that the larger it is, the better we expect the clustering to be. An (ideal) value of 1.0 would mean that for all covered basins, we obtain one useful (defined by the majority of constituents) cluster and no others. A value of 0.5 could mean a DBF of 1.0 (all basins have a useful cluster) but a UCF of 0.5 (only half of all obtained clusters are useful), the converse, or any other similar combination.

Nevertheless, both values, UCF and DBF, are pessimistic measures in the following sense. Even for "*unuseful*" clusters, the local optimization will most likely stop at *some* local optimum, only that the exact location of this optimum is largely influenced by random effects. Thus, we will most likely not need to "*evaluate*" (start a local optimization with) all determined clusters to detect the fraction of basins we can expect to obtain, DBF. But the larger the UCF, the fewer the mistakes we can make in selecting the best clusters as candidates for local optimization, and the more we differ from just randomly selecting start points as done in CMA-ES.

Task

The overall task of this experiment is to choose a suitable NBC variant per dimension, and to provide a good setting for the clustering distance ϕ and a matching range for the sample size. We will do that in a two-step procedure. At first, the best variant per dimension is chosen by means of aggregating the results of different ϕ values. For each combination of variant, dimension, and sample size, we compute the rank sums and choose the variant with the smallest rank sum as representative. In order to avoid a too complicated decision rule for applying the NBC, we apply the majority rule for the variants representing the same dimension. Our measurement will be the product of UCF and DBF, and all values should be aggregated from a number of different test problems.

In a second step, we analyze the performance of the chosen variants for the single ϕ values and generate a rule that is as simple as possible. The criteria for this experiment seem to be relatively fuzzy. However, this resembles well the problem of defining "*default values*" for a method that can be applied to a wide variety of optimization problems without actually running it on too many of them. Nevertheless, we hope that the obtained values are at least robust enough to suit most applications.

Setup phase 1

We run each of the eight variants given by the combinations with/without correction (see Sect. 4.4), LHD/uniform random initialization, and rule 2 (see Sect. 4.5), 50 times on each test problem and each combination of dimensions $\in \{2, 3, 5, 10, 20\}$ and sample size $\in \{100, 200, ..., 800\}$, and measure the average UCF and DBF values. This is repeated for every clustering distance (ϕ) parameter value $\in \{1.5, 1.6, ..., 2.5\}$. At the same time, we record the number of covered basins (the ones that contain at least one sample).

As test problems, three configurations of the MPM (single peaks) problem are chosen: MPM-10, MPM-20, and MPM-100, with 10, 20, and 100 peaks, respectively. Additionally, a funnel problem with five funnels and 100 peaks is employed, named FMPM-5/100. The parameters of these generator configurations (note that for each single run, a different instance is generated) are given in Table 4.3 or otherwise are the default values of Table 4.2. For each variant, sample size, dimension, and ϕ combination, the average UCF and DBF values are averaged and multiplied. We then compute the rank sums over all factors except ϕ.

Table 4.3 Parameter ranges for the four test problems. Single instances are generated by determining concrete values within these ranges from a uniform random distribution.

short name	funnels	funnel radii	peaks	peak radii
FMPM 5/100	5	10 − 20	100	1 − 3
MPM-10			10	5 − 10
MPM-20			20	5 − 10
MPM-100			100	1 − 3

Result phase 1

The obtained DBF and UCF values for two of the test problems are shown in Figures 4.10 and 4.11. The two other MPM-based test problems are quite similar to MPM-20 in structure and only differ in the values. Figure 4.12 gives the number of covered basins for an LHD sample; differences with random initialization are small. Figure 4.13 depicts the best variants based on the measured rank sums over the product of UCF and DBF over the dimensions and sample sizes.

Observations phase 1

The multi-criterial view on UCF and DBF values for MPM-20 (Figure 4.10) demonstrates that the search space dimension has huge influence on the performance of the single variants. Switching on and off the correction is the most important of the three factors, but correction only has clear advantages in dimensions 2 and 3, especially with large sample sizes. The variants without correction generate much larger DBF values, and from dimension 5 on, their UCF results are at least comparable to the ones applying correction. If run with a matching ϕ value, the *nocorr-ran-rule2* variant appears most suitable for most situations. In dimension 20, differences between the variants are rather small. DBF values remain too small to be of much practical use, regardless of the sample size.

For the FMPM-5/100 problem, the results displayed in Figure 4.11 are a bit more extreme. We can easily see that the variants appyling correction are much better for large sample sizes and small dimensions than the ones without. Interestingly, the latter variants get worse with increasing sample size. In dimension 20, we find variants with satisfactorily high UCF values, but the DBF values are so low that the clustering is probably not very useful (note that the problem possesses five funnels; thus a DBF of 0.4 would just be enough to clearly differentiate two of them).

The covered basins overview of Figure 4.12 shows that any sample size is enough for FMPM-5/100, and for MPM-10 up to dimension 5. For higher dimensions, the sample size has visible influence and even 800 elements are not enough to cover all basins. This is probably due to large flat areas between the single peaks. For MPM-20 and MPM-100, the largest samples seem to be good enough to cover nearly all peaks, and the smaller samples only cover a minor fraction of the basins. For dimensions 2 and 3, especially for the MPM-100, several basins seem to be unreachable and thus cannot be covered. These are most likely overlay effects of larger basins hiding small ones, which is currently not prevented by the MPM generator.

In Figure 4.13, we see that the variant without correction, with random initialization, and with rule 2 dominates for dimensions 5 and 10, and for the smaller samples of dimensions 2 and 3. For medium sample sizes in two or three dimensions, the variant with LHD initialization and without rule 2 is better, and for really large sample sizes,

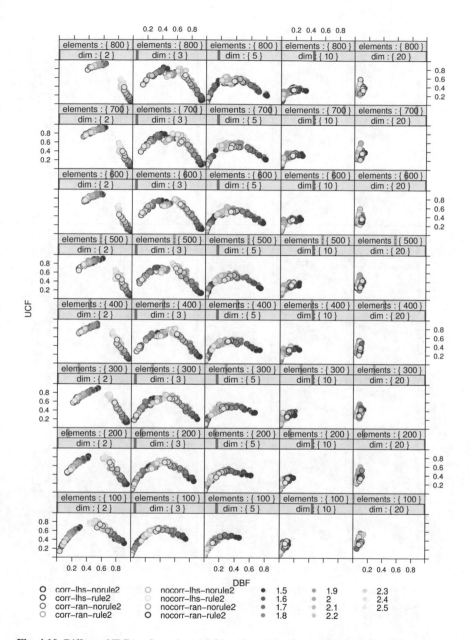

Fig. 4.10 Different NBC configurations (eight variants, 11 values of the clustering distance ϕ) over 100 to 800 elements and two, three, five, ten, and 20 dimensions. The underlying problems are 20 peaks MPM instances. The outer circle marks the method combination, the inner one the clustering distance parameter ϕ. UCF and DBF are the objectives (maximized), standing for the usefulness of detected clusters and the fraction of detected basins.

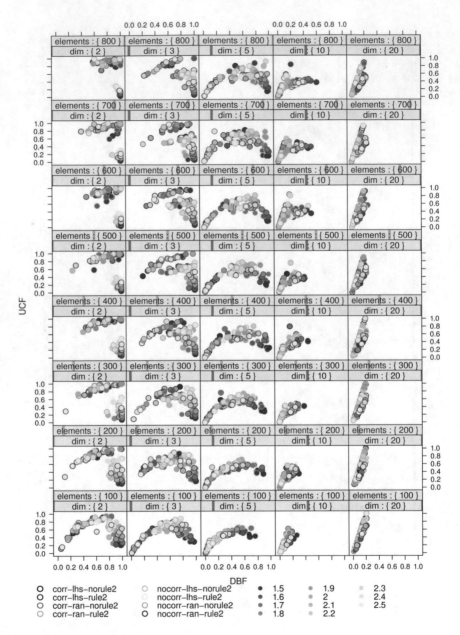

Fig. 4.11 Different NBC configurations over 100 to 800 elements and two, three, five, ten, and 20 dimensions, on a five-funnel 100 peak FMPM problem. Symbol semantics as in Fig. 4.10.

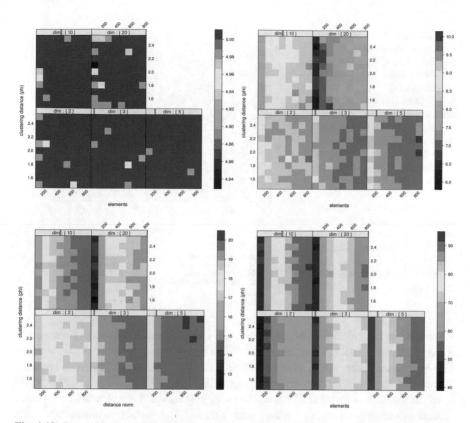

Fig. 4.12 Covered basins (with at least one sample point) for an LHD sample on the four problem types, top row from left to right: FMPM-5/100 and MPM-10. Bottom row: MPM-20 and MPM-100. Averaged over 50 repeats. Note the different scaling (see problem properties in Table 4.2). Due to the randomized size, shape, and location of peaks, some peaks seem to be hidden for MPM-100, two and three dimensions. Note that the clustering distance parameter ϕ can be safely ignored here as the clustering has no influence on the number of covered basins. However, the different rows can be regarded as repeated measurements and show the small variance of the obtained values.

correction should be switched on with random initialization with or without rule 2. The result for dimension 20 shows some diversity, probably due to the overall small values obtained. The variants applying correction form the majority over the eight sample sizes, but we cannot obtain a clear picture for recommending a single variant.

Discussion phase 1

The results for MPM-20 and FMPM-5/100 (and this also holds true for the two unshown problems) are much more similar than may be expected, and they show the

Fig. 4.13 Best NBC configurations over 100 to 800 elements and two, three, five, ten, and 20 dimensions, based on aggregation of all four test problems. Obtained by rank sum comparison over all ϕ values per element/dimension combination. The eighth variant, *corr-lhs-rule2* was not determined best for any setting. Note that the absolute values for dimension 20 are very low, so that a decision based on the rank sum values is questionable. However, for most cases, the configuration with random initialization, with rule 2 applied, and without correction seems to be most robust.

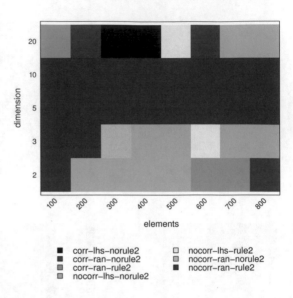

same tendencies: of all factors, the correction has the largest influence, followed by the application of rule 2, and the effect of the initialization type is relatively small. Generally, correction seems to be more advisable the smaller the dimension and the number of local optima are, and the larger the sample size is. For FMPM-5/100, it seems that the correction filters out the single peaks and enables detecting most funnels accurately (at least in dimensions 2 and 3); the variants without correction largely fail at generating useful clusters here.

We state that the decision for default parameters we are going to make now necessarily means finding a compromise. Even for similar problem types, the parameter choice would be slightly different. However, in a black-box optimization, we usually do not know much more than the dimension, so parametrization based on problem features is not possible. However, the sample (initial start population) size is algorithm-dependent and can be set according to external conditions such as the maximum number of allowed function evaluations. Thus, we need to fix default NBC variants for two factors, and except for dimension 20, it is straightforward to generate a decision rule from Figure 4.13. In this case, the absolute values are so low that differences are of the same order as the standard deviations of the measurements (and thus would not get significant in any statistical test). We would argue that applying NBC in the current form does not make much sense for dimension 20 and larger. Nevertheless, if we have to name a variant, it seems reasonable to choose *corr-ran-norule2*, as it is the most simple method that applies correction. The full decision rule is provided in Figure 4.14.

It is one weakness of this experiment that the number of test problems we can consider is relatively small, but all test problems were generated anew per run, so

default variant: no correction, random initialization, rule 2 enabled

dimension 2: for sample sizes > 100, but ≤ 500 elements: no correction, LHD initialization, rule 2 disabled

dimension 2: for sample sizes > 500: correction, random initialization, rule 2 enabled

dimension 3: for sample sizes > 300: no correction, LHD initialization, rule 2 disabled

dimension 20: correction, random initialization, rule 2 disabled

Fig. 4.14 Rule for choosing the default NBC variant according to dimension and sample size

a strong impact of single instances is prevented. Still, one can argue that the NBC is adapted here to a specific class of problems, namely those in lower search space dimensions ($D \leq 10$) and with a low to medium number of peaks. This is certainly true and intended, as this is the range of problems niching algorithms seem to be most useful for.

Another, but weaker, counter-argument is that the generator does not prevent the overlay of peaks, which becomes very visible for some parametrizations and could affect the results. However, we argue that the importance of the overlay effect is limited, as even in the worst case (dimension 2 for the MPM-100) still around 70% of the peaks were visible, and the conditions were the same for all variants. For the MPM-20, only one or two peaks are concealed for $D = 2$ (none for $D > 2$).

Setup phase 2

No new data must be generated; we only need to apply the decision rule of Figure 4.14 and display the results for the chosen variants in more detail, also including the single ϕ parameter values.

Result phase 2

Figure 4.15 shows the results for the chosen variants in the desired form.

Observations phase 2

For dimensions 2 and 3, with few exceptions, it seems favorable to apply very large ϕ values that stand for very reluctant clustering, that is, only very high relative neighborhood distances trigger splitting off a cluster. For most other cases, very small parameter values seem to be more appropriate. In this case, rather too many clusters are generated. Up to dimension 5, the best ϕ values have a tendency to increase

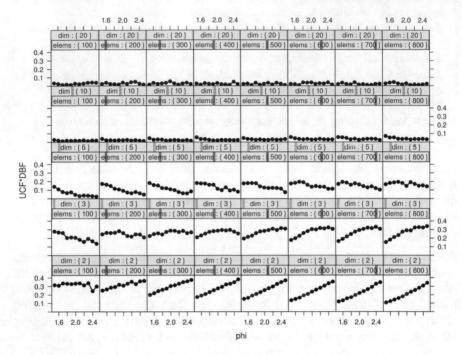

Fig. 4.15 Effect of changing ϕ values from 1.5 to 2.5 for the (per dimension/element number) selected NBC configurations over 100 to 800 elements and two, three, five, ten, and 20 dimensions, based on aggregation of all 4 test problems.

with growing sample size, which means that the NBC is forced to be more cautious about generating new clusters as more data becomes available. For dimension 20, the results do not allow for selection of proper ϕ values; no clear pattern is recognizable.

Discussion phase 2

Except for the case of dimension 20, it is not too difficult to generate a rule for choosing ϕ. The choice is supported by the observed stability and mostly only gradual shifts over the different sample sizes. The resulting rule is given in Figure 4.16. For dimension 20, the choice is rather arbitrary. We decide to go with the value that works best for dimension 10, namely $\phi = 1.5$.

default value: 1.5
dimension 2: for sample sizes $\leq$ 100: 2
dimension 2: for sample sizes $>$ 100: 2.5
dimension 3: for sample sizes $>$ 100 but $\leq$ 400: 2
dimension 3: for sample sizes $>$ 400: 2.4
dimension 5: for sample sizes $>$ 400: 1.7

Fig. 4.16 Rule for setting ϕ, the clustering distance parameter, with regard to the NBC variant chosen by means of the rule given in Figure 4.14.

4.7 Conclusions

Reviewing the performance observed in experiment 4.6.4.1, we can conclude that the NBC clustering works very well up to dimension 5, and reasonably well in dimension 10. Above that, the achieved clustering is unlikely to be of any use. This is not a problem of our method but one of all geometry-based algorithms, as already stated by Beyer et al. [32] in a data-mining context. By applying problem knowledge, it may be possible to slightly extend this range, e.g., to dimension 20, but we doubt that this can be achieved in the general case.

Nevertheless, we emphasize that a topology-based method such as ours can provide a helpful clustering before the optimization itself is started, from an initial sample. And we have seen that this sample does not need to be very large; sizes of 40D to 50D (D being the search space dimension) seem to be sufficient. More sample points do not necessarily improve the clustering, which does not really suggest applying an archive method. The provided method can thus be added to many niching or global search algorithms in order to improve the separation of the initial population into basins, or basin groups in the case of a funnel problem. The algorithmic complexity is quadratic in terms of the initial sample size due to the distance matrix computation, which is surely tolerable for most black-box optimization applications. Furthermore, NBC is also suitable for "*exotic*", non-real search spaces. The only precondition to its applicability is that, by some means, a distance matrix can be obtained that represents the differences between single search points well.

At this point, one may ask the question about how the experimental results and especially the measures UCF and DBF relate to the fundamental niching prospects investigation of Chapter 3. There, we have been arguing on a different level, that of clusters and local searches. Here, we have a real search space with individual search points that form a cluster. Some effects can occur only if single search points are considered; for example, mixed clusters cannot be represented in the groundwork model. Furthermore, the two models work on different time scales. Regardless of whether the groundwork model employs several subsequent clusterings to detect how long it takes to obtain the global optimum or all optima (t_2 or t_3, respectively), we only deal with one time step in the search point model.

Transferring the results from the search space model back to the groundwork model would not be possible without establishing many definitions (for example, about how the actual clusterings in the search point model look). However, it is also clear that a high probability of detecting points in the same basin, $p_{BI}(\mathbf{x}_1, \mathbf{x}_2)$, is helpful for achieving a high number of useful clusters. What we can do to support the connection between the models is to measure UCF and DBF values that would be obtained under different basin identification probabilities. Note that the results given in Table 4.4 can be considered as marginal, as many important effects are ignored in the groundwork model. Measured values in the search point model can be expected to be much lower, but to follow the same trend.

Table 4.4 UCF and DBF values for an abstract problem as treated in the groundwork model in Chapter 3, under different basin identification probabilities p_{BI}. Equal probabilities of hitting each basin are assumed. Note that the simplified model does not represent all important effects and thus real values would be lower. Numbers are averaged from 10,000 repeats. For $p_{BI} = 1$, UCF is always 1.

basins	clusters	p_{BI}	UCF	DBF
10	10	0	0.65	0.65
10	10	0.5	0.85	0.85
10	10	1.0	1.00	1.00
20	10	0	0.80	0.40
20	10	0.5	0.91	0.45
20	10	1.0	1.00	0.5

Chapter 5
Niching Methods and Multimodal Optimization Performance

What is the concrete task of multimodal optimization methods and how can we compare the performance of these algorithms experimentally in a meaningful way? Which niching techniques are actually applied in existing algorithms, and how do they relate to the nearest-better clustering method? Specific benchmark suites and performance measures are still evolving and far from being mature. We review the current state and envision future developments in this respect.

The main focus of this chapter is to provide all constituents of an experimental evaluation of the optimization algorithms we are going to set up based on the nearest-better clustering method described in Chapter 4. Surprisingly, published works in the area of evolutionary multimodal optimization have been sorted into folders based on to their historic roots (such as real-valued *genetic algorithms*, GAs, *evolution strategies*, ESs, *differential evolution*, DE, or *particle swarm optimization*, PSO) for a very long time, so the developments of the other branches have often gone unnoticed. In every single subfield, researchers were used to comparing innovative approaches to the archetypal works of Richardson and Goldberg [91] (sharing), De Jong [56] (crowding), Mahfoud [137] (niching), and Pétrowski [166] (clearing). This is probably the main reason that necessary developments were prevented for quite some time, from concrete target definitions to survey papers to competitions. In evolutionary global optimization (a single global optimum to be found in the shortest time), the *black box optimization benchmarking* (BBOB) definitions are currently getting established as a quasi-standard [101] for the comparison of algorithms. These comprise a reasonably composed set of benchmarking problems, a well-defined experimental setup, and a procedure for performance measuring and visual comparison. Unfortunately, we are still far from that situation for multimodal optimization. Besides the missing communication between different branches of EC, a reason for this may be that transferring the paradigm of natural evolution onto multimodal optimization is not trivial. Interestingly, there is no single instance in nature that pursues the search for several local or global optima at once.

In Sect. 1.2, we outlined our view on multimodal optimization, but without providing a concrete performance criterion. In fact, choosing such a measure is not trivial.

115

Preuss and Wessing [180] give an overview of currently employed measures and the ones that can easily be derived from them. As we will see, possible measures can have tight connections to niching and selection mechanisms as used in multimodal and also in multiobjective optimization. The example of the dominated hypervolume ($\mathscr{S}$-metric) shows that this is not an unusual situation in EC.

Interestingly, the term *multimodal optimization* itself has not been very common until very recently, being used occasionally when it was the main task of the optimization process to come up with several instead of only one good solution. But which ones, and how many? As of 2015, there still exists a certain amount of fuzziness around targets and definitions for this research area, and only in 2011 did Das, Maity, Qu, and Suganthan provide the first survey paper [55] that highly dwells on different niching techniques and the optimization algorithms that employ them. They state:

"*multimodal optimization refers to the task of finding multiple optimal solutions and not just one single optimum, as it is done in a typical optimization study.*"

(Das et al. [55], page 1)

It goes without saying that such endeavors only make sense if the tackled problems are multimodal, but it is widely assumed that most application problems that cannot be reliably solved by means of classic optimization algorithms (e.g., gradient and quasi-Newton methods) and are thus tackled by metaheuristics are of this nature. How does *niching* relate to this task? We already provided some definitions of niching in Sect. 3.2, including our own, introduced in [168]. Das et al. refer to niching as techniques rather than algorithms, so that in principle any niching technique could be added to any EA:

"*Numerous evolutionary optimization techniques have been developed since [the] late 1970s for locating multiple optima (global or local). These techniques are commonly referred to as* niching *methods.*"

(Das et al. [55], Abstract)

We share this view to a large extent and will therefore in the following start by deriving a number of concrete use cases for multimodal optimization that lead to a list of suitable performance measures in a second step. This will be followed by a catalogue of currently employed niching techniques. Note that it is in some cases difficult to separate optimization algorithms from the niching methods they utilize.

5.1 Use Cases

What different cases do we have to take into account for setting up performance measures?

Niching methods may be targeted at finding all global optima of an optimization problem, or the global and some good local optima, or even all optima. In a benchmark situation with constructed test problems, the locations of all optima are usually available or can be computed, and it is possible to record these locations also for test problem generators (Yuan and Gallagher [249, 85], Preuss and Lasarczyk [173]). If we want to compare performance on real-world problems, this information is mostly missing; one may have just the location and objective values of some previously detected optima available, and no clue about the global optimum.

However, even in a benchmark situation, the question of what to look for cannot be easily answered if there are many global or local optima. It may be an interesting and challenging task to detect all 500 global optima of a benchmark problem (Deb and Saha [60]), but we doubt that this is very meaningful for a real-world application. It seems to be accepted among most researchers that in these cases only a subset of the optima should be pursued, namely one that provides a good trade-off between diversity in search space and the best attainable objective values. Note that these are several requirements, and the degree of conflict between the different objectives highly depends on the tackled problem. Preuss and Wessing [180] therefore recommend applying tools from the domain of *evolutionary multiobjective optimzation* (EMO), but no consensus has been reached yet on the concrete measures to employ. The reasoning behind demanding search space diversity is, however, clear. In a real-world application, the decision maker shall be provided with a (low) number of alternatives of high quality because, for the implementation of these solutions, problem knowledge may be available that has not been coded into the optimization problem and that may rule out some but not all of them.

Taking into account all of the above, we may set up three different categories:

All-Global Only the (usually) low number of known global optima are pursued. The local optima are not known or are considered unimportant. All of the problems of the CEC 2013 niching competition [129] belong to this category.

All-Known All optima, global or local, have to be found. The six-hump camel back (also contained in [129], but with a different task) is a typical test problem in this respect.

Good-Subset Some optima have to be found, and they should spread well over the search space. The appropriate number of targeted optima and the desired diversity measure have to be specified.

Note that the first two categories are only suitable for benchmark problems because the locations of the optima must be known. In order to be complete, we may add a fourth category that comprises the global search problems as considered in BBOB [101]:

One-Global One global optimum has to be found as quickly as possible (usually measured in function evaluations),[1]

---

[1] Note that this is based on the assumption that function evaluations need almost constant time without respect to what is evaluated. This may be wrong, e.g., the calculation of constraints that render a solution invalid can be much quicker than a full evaluation.

While for the first three categories, run times are usually ignored (except an upper limit of allowed function evaluations), this is the primary criterion for measuring in the BBOB context. The reasoning for preferring the target-oriented measuring over the budget-oriented one is simply that the obtained values can be much better interpreted in the first case. An algorithm that reaches the same objective value in half the time is obviously twice as fast. In the second case, the scale of the different objective values is unclear. While we still have the ranking information (smaller is better for minimization), we cannot know how much better we are if we can decrease the objective value by 10%. This is clearly problem dependent.

For *all-global* and *all-known*, it would be very simple to define targets (all desired optima have to be found with a certain precision), and one could obtain a good overview of the abilities of different algorithms by setting up multiple targets (in the BBOB, a set of around 50 targets is used). We only have to make sure that the functions are not too difficult, so that the tested optimization algorithms are at least sometimes able to reach some of the targets. In this case, the *expected running time* (going back to the ENES in Price [181], and introduced as success performance in Auger and Hansen [13]) can be employed as a measure:

$$\mathrm{ERT}(f_{\mathrm{target}}) = \mathrm{RT}_S + \frac{1 - p_s}{p_s} \mathrm{RT}_{\mathrm{US}} . \tag{5.1}$$

In this formulation, the ERT depends on the demanded target value f_{target} and the associated success rate p_s, and the running time (in function evaluations) RT_S of the successful trials. A trial is successful if the specified target is reached. However, it also depends on the running time $\mathrm{RT}_{\mathrm{US}}$ of unsuccessful trials, which are usually stopped after using up a defined maximum number of function evaluations. With different targets, such as the number of optima to be obtained, the ERT could also be employed in a niching context, but this is currently not done.

However, for the *good-subset* case, simply defining objective-based target values is not enough, we also have to take the diversity into account. For employing a measure such as the ERT, one would need to either demand fixed target values for the objective *and* the diversity, or establish some aggregation and define targets for the aggregated value. Furthermore, the selected subset should not be too large. The basic motivation for multimodal optimization itself is to provide a number of alternative solutions to the decision maker. And it is clear that the solutions we pass on to a decision maker should somehow influence our performance measure. In contrast to the situation for the *all-** categories, we now have a subset selection problem: in the former case, we simply check for the solutions that have the minimal distance to the desired optima; all others are disregarded. But what subset do we choose to base a measure on if the set of optima that shall be pursued is not small? And even worse, how could we possibly measure if the optima are not even known, as usually happens for real-world applications? Based on the collection provided in [180], we now provide an overview of available quality indicators for measuring multimodal optimization performance. Some of these have often been employed in published

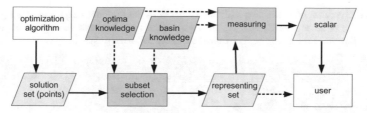

Fig. 5.1 Data flow from the solution set obtained by an optimization algorithm up to the final scalar measure value. Problem knowledge is an optional input for the subset selection and/or measuring part.

works, others are new developments. We also suggest an aggregation-based measure with a heuristic subset selection (R5S) that may be used also for real-world problems for which the optima are not known.

5.2 Available Performance Measures

What measures are available for the different use cases, and what are their advantages and disadvantages?

We expect any multimodal optimization algorithm to provide a solution set, not a single solution, after termination. The cardinality of this set is not necessarily restricted (except for an imposed maximum number of function evaluations); it could be the final population or an archive. Additionally, in generated benchmark functions, the locations of the sought optima and/or of the basins these optima reside in may be available. Whether or not it is possible to compute diversity measures or simple statistics over the whole solution set, in all other cases some kind of subset selection has to be applied before a measure can actually be computed. As an example, the popular *peak ratio* that is applied in [129] requires for every sought optimum to select the solution from the set that possesses the minimal distance to it. We might therefore describe the process of measuring as a data flow that is depicted in Figure 5.1. We term the selected subset a *representing set*, as all the other solutions contained in the original solution set have no influence on the result of the measuring. The final measure is then obtained by applying a quality indicator to the representing set. Table 5.1 provides an overview of indicators, taken from [180], and their requirements regarding objective value, optima, and basin information availability. Some of these indicators have been analyzed by Emmerich et al. [75] in the context of level set approximations, with an application to multiobjective optimization in mind. However, as already stated, the application to multimodal optimization is reasonable and straightforward. Another idea from multimodal optimization, namely to compute diversity from search space location and objective values as suggested in Shir et al. [211], is not taken into consideration here as in MMO there is only one objective

Table 5.1 Overview of available quality indicators with their requirements. *requires* $f(\mathbf{x})$ means that objective values are used, *subset select.* that a method must be prepended to filter out unwanted solutions, *param.* that a parameter is required that may be problem-specific.

indicator	short	requires $f(\mathbf{x})$	subset select.	optima known	basins known	param.
sum of distances	SD					
SD to nearest neighbor	SDNN					
Solow-Polasky diversity	SPD					✓
average objective value	AOV	✓				
peak ratio	PR		✓	✓		✓
quantity-adjusted PR	QAPR			✓		✓
peak distance	PD		✓	✓		
augmented PD	APD	✓	✓	✓		
peak accuracy	PA	✓	✓	✓		
averaged Hausdorff distance	AHD			✓		✓
augmented AHD	AAHD	✓		✓		✓
basin ratio	BR		✓	✓	✓	
quantity-adjusted BR	QABR			✓	✓	
basin accuracy	BA	✓	✓	✓	✓	
representative 5 selection	R5S	✓				

value. In the following, we simply list the different indicators (grouped according to their requirements) with a short description. For the more popular measures, we also provide equations, where the set of optima is denoted by $\mathcal{Q}$ and the approximation set is denoted by $\mathcal{P}$. The distance of an element $\mathbf{x} \in \mathcal{P}$ to its nearest neighbor is then

$$d_{\mathrm{nn}}(\mathbf{x}, \mathcal{P}) = \min\{\mathrm{dist}(\mathbf{x}, \mathbf{y}) \mid \mathbf{y} \in \mathcal{P} \setminus \mathbf{x}\} . \tag{5.2}$$

More details can be found in [180].

Note that instead of completely relying on the subset selection from an archive or population being done during the measuring process as described here, one could also expect that the optimization algorithms do this at least partly on their own in order to obtain a reasonably small result set. This would certainly be an improvement over the current state, as most algorithms for multimodal optimization currently ignore this requirement. In the following, we also discuss some new *quantity-adjusted* measures that penalize large result sets, and employing these could be an incentive to move into this direction.

5.2.1 Indicators That Require No Problem Knowledge

These indicators may of course be applied for benchmark and real-world optimization as they do not need any additional information besides the solution set itself. However, they are of limited use for drawing conclusions about the state of the optimization.

Sum of Distances (SD)

The sum of distances simply sums up the pairwise distances of all elements of a population (in the search space). A reasonable magnitude of the result can be obtained by applying the square root.

$$SD(\mathscr{P}) := \sqrt{\sum_{\mathbf{x}\in\mathscr{P}}\sum_{\mathbf{y}\in\mathscr{P},\mathbf{y}\neq\mathbf{x}} \text{dist}(\mathbf{x},\mathbf{y})} \ . \tag{5.3}$$

Sum of Distances to Nearest Neighbor (SDNN)

In comparison to the SD indicator, the sum of distances to the nearest neighbor penalizes the clustering of solutions, as only one (namely the smallest) distance is considered per solution.

$$SDNN(\mathscr{P}) := \sum_{\mathbf{x}\in\mathscr{P}} d_{nn}(\mathbf{x},\mathscr{P}) \ . \tag{5.4}$$

Solow-Polasky Diversity (SPD)

Solow and Polasky [214] developed this indicator for measuring the diversity of a biological population and showed that it possesses several properties that the sum of distances indicator misses. As an example, the indicator value does not change if an already contained solution is added a second time. It has been introduced into multiobjective optimization by Ulrich et al. [234]. Unfortunately, the obtained value is critically dependent on a good setting of one parameter.

Average Objective Value (AOV)

As a simple population statistic, the average objective value could be used to rank the quality of different solution sets.

5.2.2 Indicators That Require Optima Knowledge

The following indicators use the locations of the sought optima and are thus only applicable in a benchmark scenario. Note that this information is not even available for all benchmark problems. Either it has to be provided explicitly at construction time or the optima may be found by subsequent local searches.

Peak Ratio (PR)

This is probably the most employed indicator for multimodal optimization. However, it is being used in several versions that mainly differ in the criterion for detecting that a peak has been found. It has been suggested by Ursem [236] and requires the definition of a range constant r in order to decide if an optimum has been approximated well enough in the search space. The version employed in [129] also uses an accuracy constant ε to make sure that the objective space distance is smaller than a predefined accuracy. Regardless of the detection criterion, the indicator is then computed as a fraction of the numbers of found and sought optima. Let k denote the number of sought optima, then the Peak Ratio (simplified Ursem version without crowd counter) can be computed as:

$$\mathrm{PR}(\mathscr{P}) := \frac{|\{\mathbf{z} \in \mathscr{Q} \mid d_{\mathrm{nn}}(\mathbf{z}, \mathscr{P}) \le r\}|}{k}. \tag{5.5}$$

The principal weakness of this indicator is its discrete nature. A peak is found or not found, there is nothing inbetween (any gradual change cannot be detected until a threshold is crossed).

Quantity-Adjusted Peak Ratio (QAPR)

The peak ratio indicator does not penalize large solution sets (the subset selection just returns the solutions that are nearest to each sought optimum). This may be amended by dividing not by the number of optima but by the square root of the product of the number of optima and the solution set size, as suggested in [180]. This indicator does not need an explicit subset selection because optimization algorithms are incited to produce small solution sets.

Peak Distance (PD)

This indicator is related to the peak ratio, but uses the minimal distances of solutions to sought optima directly, without a threshold distance. It has been introduced as "distance accuracy" by Stoean et al. [221]. We use it in the form suggested in [180], where the sum of minimal distances is divided by the number of sought optima. This addition relates peak distance to the indicator *inverted generational distance* [51], which is used in multiobjective optimization to compute the distance to a desired set. Again, $\mathscr{P}$ means the current population, and the $\{z_1, z_2, \ldots z_k\}$ form the set of sought optima $\mathscr{Q}$.

$$\text{PD}(\mathscr{P}) := \frac{1}{k} \sum_{i=1}^{k} d_{\text{nn}}(z_i, \mathscr{P}) . \tag{5.6}$$

Peak Accuracy (PA)

Instead of search space distances, the peak accuracy considers the sum of differences of the objective values of the elements in the solution set that are nearest to the sought optima. It was introduced by Thomsen [225], and [180] suggests dividing by the number of sought optima to make it consistent with PR and PD, and renaming it to *peak inaccuracy*. As stated in [221], peak accuracy can lead to misleading results because one element of the solution set may be the nearest neighbor for several optima in which case the search space distances (which may be reasonably large) are ignored. Of course, it shares this problem with the peak ratio measure. We take over the equation from [180] here, with $\text{nn}(z_i, \mathscr{P})$ meaning the nearest neighbor of z_i in $\mathscr{P}$.

$$\text{PA}(\mathscr{P}) := \frac{1}{k} \sum_{i=1}^{k} |f(z_i) - f(\text{nn}(z_i, \mathscr{P}))| . \tag{5.7}$$

This indicator is closely related to the *maximum peak ratio* (MPR) indicator as introduced by Miller and Shaw [151] and heavily employed by Shir [207, 208].

Averaged Hausdorff Distance (AHD)

This indicator was defined by Schütze et al. [201] and is related to the inverted generational distance and thereby also to the peak distance indicator. The idea is to measure the distance to a target set. The AHD contains a parameter p in order to control the influence of outliers. We set $p = 1$ as in [75] here, which simplifies the equation to:

$$\text{AHD}(\mathscr{P}) := \max \left\{ \frac{1}{k} \sum_{i=1}^{k} d_{\text{nn}}(z_i, \mathscr{P}), \frac{1}{\mu} \sum_{i=1}^{\mu} d_{\text{nn}}(x_i, \mathscr{Q}) \right\} . \tag{5.8}$$

The AHD differs substantially from the previous indicators as it penalizes unnecessary points that are far from any sought optimum (if applied in the multimodal optimization context). The indicator was originally envisioned for multiobjective optimization.

Augmented Averaged Hausdorff Distance (AAHD) and Augmented Peak Distance
(APD)

Both AHD and PD measure only in the search space and disregard the objective
values. This leads to the undesired effect that two points with the same distance to a
sought optimum are valued equally although their objective values can vary highly.
In order to counter this problem, [180] suggests augmenting the indicators by simply
appending the objective values as additional components to the search space location.
Scaling issues can be treated by utilizing the objective values of the sought optima
for normalization.

5.2.3 Indicators That Require Basin Knowledge

Once an optimization algorithm has reached a certain basin, its associated optimum
can be obtained simply by executing a local search. As evolutionary algorithms have
a tendency to lose covered optima rather than to find new ones (due to the genetic drift
effect, as investigated in Preuss, Schönemann, and Emmerich [177]), it makes sense
to evaluate basin coverage as a way to estimate the number of obtainable optima
very early in the optimization process. However, this is of course not applicable
in a real-world problem, and even for benchmarks we have to rely on additional
information in order to check if a solution is located within a certain basin or not.
For most benchmark problems, this information is not available, but it may be coded
into generators or detected by subsequent local searches in the same way as for the
optima (Sect. 5.2.2). The following indicators have been suggested in [180].

Basin Ratio (BR)

In analogy to the peak ratio, the basin ratio is computed by counting the number of
covered basins (by checking for each basin if it contains at least one point of the
solution set) and dividing this by the total number of basins. This indicator could be
interesting for checking if and how fast basins are lost throughout the optimization
process.

Quantity-Adjusted Basin Ratio (QABR)

This is the counterpart of the QAPR measure, formulated for basin coverage. Instead
of dividing by the number of basins, we take the square root of the number of basins
multiplied by the solution set size in order to penalize large solution sets.

Basin Accuracy (BA)

Basin ratio and peak accuracy can be combined by replacing the binary count in BR with the minimal objective value difference to the associated optimum for all solutions within a certain basin. If a basin is not covered by any solution, a penalty is applied. Preuss and Wessing [180] suggest to use the maximum possible objective space difference. The sum of these values is again divided by the number of basins.

5.2.4 A Measure for Real-World Problems: R5S

The *representative 5 selection* (R5S) was suggested in [180] in order to provide a subset selection method and, following from that, a measure for real-world situations, where neither optima nor basins are known and only a small number of very good solutions is desired. The number 5 is often given by practitioners as orientation for the expected solution set size, and the method is designed to select around five diverse but very good solutions, even if this is not explicitly coded into it. It rarely selects more than seven solutions, but fewer of course if the original set is too small or has little diversity. This resembles the *good-subset* use case of Sect. 5.1. The heuristic nature of the R5S method only guarantees that at least two solutions are returned (if more than one was provided), because the second solution always improves the diversity of the selected set.

Algorithm 4: Representative 5 selection (R5S)

1 compute all search points' mutual distances;
2 **forall the** *search points* **do**
3 $\quad\lfloor$ find nearest search point and nearest point that is better, record distances;

4 sort points after objective values from best to worst;
5 **forall the** *search points* **do**
6 $\quad\lfloor$ $ds(point)$ = sum of distances to all worse points with weight $1/2^i$,
7 $\quad\lfloor$ for $i = 1 : |worse\ points|$;

8 **if** $|search\ points| > 2$ **then**   // remove probably uninteresting points
9 $\quad\lfloor$ delete points where nearest neighbor distance equals nearest better distance

   // remove dominated points
10 **forall the** *search points in reverse order up to third best* **do**
11 $\quad\lfloor$ remove point if better point with higher $ds(point)$ exists;

12 return remaining points;

Algorithm 4 provides a pseudocode description of the diversity computation and selection process. It starts by detecting the nearest and the nearest better solution for each element of the solution set. These are used later for removing solutions that are with a high probability not interesting because if the nearest neighbor is also the nearest better neigbhor, then the considered solution is most likely not adding much diversity to the selected set and the better neighbor is going to be chosen first (we sort by objective values in the next step). In the second `forall` loop, the relative diversity ds(point) of every solution is determined by computing a weighted sum of distances to all worse solutions, with the weight decreasing exponentially. This guarantees that the sum can never be larger than the largest possible search space distance. The distances to better solutions are not considered because they have already been taken into account while computing the relative diversity for these. In the last step, we perform a non-dominated sorting of the solutions according to objective values and relative diversities, which can be established with a single `for` loop because the solutions are already sorted by their objective values.

As we finally end up with a small subset, an objective value, and a diversity measure for every element, it is straightforward to compute a numerical value by determining the dominated hypervolume of this set. This number can be used to compare the quality of solution sets provided by different algorithms on every possible test problem. Furthermore, it is also bounded, as the hypervolume cannot get larger than the largest possible objective value difference multiplied by the largest possible

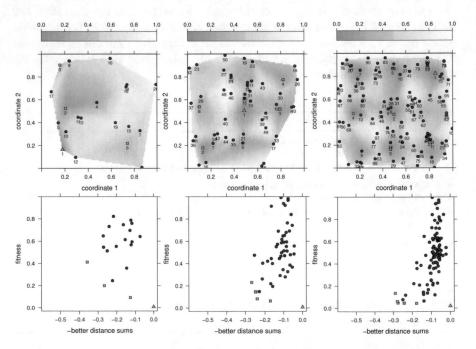

Fig. 5.2 Top: Example of R5S chosen solutions for 20, 50, and 100 randomly placed points on the cosinus problem, with an interpolation of the fitness landscape in the background. Bottom: The chosen solutions in the objective value/relative diversity space. The numbers give the objective value ranks; the triangle marks the best solution that is always chosen; the points denoted by squares are also selected.

search space distance. Note that it is not necessary to compute the former exactly; we only need to define some limits. However, the measure also possesses a peculiarity: The objective value of the best solution is not directly taken into account because its diversity value is zero. Nevertheless, we presume that this is not absolutely necessary because we strive for a good set measure, and the objective values of all subsequent selected solutions do count.

For two-dimensional problems, a visualization can easily be obtained in order to check if the subset selection indeed works as intended. We have provided an example in [180] with the simple separable cosinus function $f_3(\mathbf{x}) = \sum_{i=1}^{D} \cos(6\pi x_i)$. This problem possesses 3^D peaks, which are all globally optimal. Figure 5.2 shows the selected solutions from sets of 20, 50, and 100 random samples in 2D. As expected, R5S often chooses top-ranked solutions, but disregards them if they are located too near to already selected ones. For other simple multimodal test functions (not shown here), we have obtained very similar results. The number of selected solutions is remarkably stable over sample sizes and in search spaces of different dimensions up to 20D. Beyond that, it is probably not very meaningful to argue with geometry-based measures as all distances get more and more similar.

5.3 Niching Techniques Overview

How can we arrange the niching techniques found in optimization algorithms from different research areas into a suitable taxonomy?

At the core of niching, we find separation. This is the essence of the biological definitions (Sect. 3.1) as well as those utilized in evolutionary optimization (Sect. 3.2). But how is the separation attained in existing optimization algorithms? Or, put differently, what niching techniques are known and how do they relate to each other? We employ the taxonomy suggested in Sect. 1.3.4 and review literature from the global optimization and evolutionary computation fields, largely ignoring heritage and details of the algorithms as we are only interested in the means they apply to distribute several independent searches over the search space.

One could have the somewhat naïve hope that simple evolutionary algorithms with their global search capabilities would sufficiently cope with multimodal problems. Actually, the results of Sect. 3 show that a local search procedure with a simple random restart mechanism is not that bad if we strive to detect only one global optimizer in a multimodal setting. This is essentially what the CMA-ES (Hansen and Ostermeier [103]), one of the most successful black-box optimization algorithms currently available, tries to achieve with its different variants (IPOP-CMA-ES: Auger and Hansen [14], BIPOP-CMA-ES: Hansen [99]). However, we cannot expect that a simple EA without special measures will reasonably well approximate several optima during one run (before being restarted). Without counteractions, it will soon focus on the region appearing as the most profitable. Among others, Schönemann et al. [198] show how fast this happens for a naïve evolution strategy, even without recombination. If recombination is added, the clustering tendency is even stronger, as shown by Preuss et al. [177].

While structuring methods for evolutionary multimodal optimization, Eiben and Smith [74] suggest deciding between implicit and explicit diversity maintenance. However, with regard to niching, it is not enough to boost diversity, although that may be helpful. But diversity measures do not depend in any way on the tackled problem. Niching is rather about avoiding two fundamental mistakes while exploring the search space:

"Type I Error, Local search will be repeated in some region of attraction.

Type II Error, Local search will not start in some region of attraction even if a sample point has been located in that region of attraction."

(Ali and Storey [8])

In dealing with global optimization, the authors surely did not have evolutionary methods in mind and thus do not use the term *niching*. Additionally, one does not have to rely on a local search method as it may be possible to encode the local search behavior into the main algorithm. Furthermore, concerning the type II error, we may add that it should also be a task for a niching method to actually cover most of the basins, but this clearly depends on the available budget of function evaluations. Arguing rather from an evolutionary optimization perspective, Pétrowski [167] uses the denotation: *"explicit and implicit speciation"*. We go one step further and ask for explicit and implicit basin identification.

The techniques found in algorithms for multimodal optimization may be partitioned into three main approaches, sorted in decreasing order of the extent to which they could be called niching methods according to the above statements.

A : Explicit Basin Identification establishes some kind of mapping from search space to basins that in principle allows us to determine the basin any location in the search space belongs to.

B : Basin Avoidance could also be called implicit basin identification or *basin recognition* as introduced in Sect. 3.2. We strive to avoid searching in locations we have already explored.

C : Diversity Maintenance is concerned with spreading the search over the whole available search space, without taking the search space topology into account. We also consider approaches here that slow down information exchange within populations, e.g., by introducing subpopulations or mating restrictions, but without an explicit relation of subpopulations to basins.

It may appear paradoxical that the sequential niching algorithm suggested by Beasley et al. [26] is not really a niching method according to this scheme, as its basic mechanism performs basin avoidance. However, this corresponds well with the separation displayed in Figure 1.5: According to the property scheme of Sect. 1.3.4, sequential niching does not emphasize basin identification; it parallelizes in time and employs subsequentially modified (*"derated"*) objective functions to prevent revisiting an already searched area. This reminds one of the tunneling method of Gomez and Levy [93] (a parallelized variant has been suggested in [92]). However, all methods that modify the objective function have the systematic disadvantage that in order to eliminate exactly the influence of one basin, the shape of this basin has to be known quite accurately, which is usually not the case.

In the following, we review a number of popular *"classic"* evolutionary approaches for multimodal optimization in order to put them into one of the classes defined above, and then focus on explicit basin identification methods. Note that the design goals of the algorithms may have been different, usually *one-global* for the global optimization algorithms, and *all-global* or *all-known* for the evolutionary methods. However, we will ignore this difference for the time being as it does not necessarily have an influence on the basin identification scheme. The same scheme could often be used within a very different EA or even a global optimization algorithm. Table 5.2 summarizes the basic properties of the methods reviewed in the following, with a focus on true niching (class A) techniques.

5.3.1 The Evolutionary Niching Heritage Methods

Most lists of evolutionary niching methods start with crowding by De Jong [56] and sharing by Goldberg and Richardson [91]. However, both methods do not explicitly identify basins and are thus not really niching methods according to the scheme presented above. While crowding is a local selection that essentially results in parallel hillclimbers within one population, sharing is a basin avoidance scheme that works by penalizing solutions for clustering together in good regions. A completely different path is explored by means of the tagging approach by Spears [215]: The individuals of a population are put into randomly determined subpopulations, marked with tag bits. These tag bits are evolved along with the rest of the genome, and so are the subpopulations.

The sharing idea was later extended by means of clustering techniques later, and several methods that fall into the explicit basin identification class were suggested in the following, probably starting with Yin and Germay [248]. These will be tracked in Sect. 5.3.3. A completely different branch of methods co-evolved with the quickly developing parallel hardware of the last two decades, namely the *parallel evolutionary algorithms*. Alba and Tomassini provide an overview in [7]. De Jong [57] uses the terms *coarse-grained* and *fine-grained* EA for the two fundamental approaches, parallelization on the subpopulation and individual level, respectively. The former have also been termed *island models* (see Martin et al. [138] for a more detailed discussion), the latter are also named *diffusion models* or *cellular evolutionary algorithms*; a recent summary is given in Alba and Dorronsoro [6]. An interesting variation of an island model has been suggested by Oppacher and Wineberg [161] with the *shifting balance* GA. It was originally conceived for dynamic problems, but possesses a very useful control mechanism for preventing overlaps between the search space locations actually visited by the different islands.

One may state that conceptually, both branches of parallel EA are descendants of crowding rather than of sharing, as they implement many different ways of slowing down information exchange in order to enhance diversity. With respect to the main classes defined above, all these algorithms would belong to classes *B* or *C*. They do not contain true niching methods although they may have been inspired by niching in nature.

A modern approach to diversity maintenance evolutionary algorithms is delivered by Ulrich et al. [235]. The NOAH algorithm uses diversity as an additional selection criterion and also provides an efficient subset selection method to obtain a small result set based on the Solow-Polasky measure [214].

5.3.2 Cluster Analysis in Global Optimization

A uniform randomly sampled set of points is not expected to exhibit a clear separation into clusters. If clusters are found at all, they show up by chance or as artifacts of random number generation and not because they hint at a specific basin of attraction. So how do we derive a meaningful clustering from the population, that takes the basin structure into account? In evolutionary computation, many researchers have argued that individual search points are "*pulled*" into regions of better and better quality or *high performance regions* (Parmee [165]). Thus, we can start a cluster analysis after some generations of computing. However, this is costly, because we need to perform a lot of function evaluations before we get a first grouping. Additionally, it may happen that during this *concentration* process, several important basins are lost.

In the global optimization community, clustering approaches have been integrated into the optimization algorithms from the 1970s on, and the necessary preprocessing (concentration) was at first done by deleting a fixed fraction of the worst initial samples. Translated back to evolutionary algorithms, this would be equivalent to running exactly one generation and then starting the cluster analysis.

According to Törn and Žilinskas [229], the first application of cluster analysis within a global optimization algorithm was suggested by Becker and Lago [27]. Törn [228] himself suggested a similar but much simpler algorithm only some years later, according to his own statement without knowledge of [27]. Both algorithms do concentration by removal of the worst search points of the initial sample and then apply density-based clustering techniques. The motivation for clustering here is to prevent type I errors: "... *a cluster analysis algorithm is used to prevent redetermination of already known local minima*" (Törn and Žilinskas [229]). Instead of density-based clustering, several distance-based clustering methods have been tried; Rinnooy Kan,

Boender, and Timmer give an overview in [189]. Timmer [226] introduced a single linkage method together with a theoretically motivated critical distance. Points within this distance from another point that has already been assigned to one cluster (which originally starts with one seed point) are put into the same cluster. The cluster seeds themselves are taken from the best initial samples, in worsening objective value order. In Timmer [226], an improved procedure named *multi-level single linkage* (MLSL) is introduced that makes better use of the objective values (as [189] concludes), which is basically achieved by looking at all search points of the initial sample in order of worsening objective value.

"Since the function values are used in an explicit way in the clustering process, it is no longer necessary to reduce the sample."

(Rinnooy Kan, Boender, and Timmer [189], page 19)

We assume that this is a very important qualitative change in use of the available information; it seems that the *dynamic peak identification* (DPI) by Miller and Shaw [151] and the hill-valley detection method of Ursem [236] were anticipated by Timmer. For determining the critical distance, a constant σ has to be determined. A small value $\sigma \approx 2$ leads to detecting the global optimum with high probability, but also to a consumption of too many function evaluations; a higher value reduces the effort but at the cost of reduced probability. Note that in [189], a local search procedure based on derivatives is employed, but the method could as well be run with a direct search method instead.

Another qualitative improvement is suggested by the *topographical global optimization* (TGO) method of Törn and Viitanen [230]. Instead of using a radius-based basin identification, a graph structure (the *k-topograph*) is employed. This graph is built from the initial sample by connecting each point to all of its k neighbors that have worse function values. The points remaining without incoming connections are assumed to represent local optima and are utilized as starting points for a local search procedure. This procedure uses distances as well as objective values in a way that is relatively similar to the *nearest better clustering* (NBC) method suggested in Chapter 4. However, the authors employ much higher values of k, which is usually set to $k \geq 8$. The number of potential basins that may be identified this way is therefore quite limited. Note that rule 2 of NBC works in almost the same way but k is set to 3 only. The TGO method has been reformulated for an existing set of sampled points in Törn and Viitanen [231] and tried with an iterative sampling scheme in Törn and Viitanen [232]. The term iterative, however, only refers to the initial sampling itself; the local optimization is still appended as a single last step. Ali and Storey [8] provide an overview and experimental comparison of many of the algorithms presented in this section.

5.3.3 Explicit Basin Identification in Evolutionary Algorithms

In the following, we review several class A approaches from within evolutionary computation, roughly following a chronological ordering. However, we sometimes deviate from it to show the similarities of methods or to emphasize that an already existing method has been extended. Note that even though we restrict the review to methods that perform explicit basin identification only, the number of relevant works is quite large. For this reason, we completely disregard all algorithms that fall out of this scope. A more general overview is given in [55].

The first known clustering-based approach in evolutionary computation that attempted to identify basins without providing a problem-specific niche radius was the *adaptive clustering algorithm* (ACA) by Yin and Germay [248]. It relies on k-means clustering and is thus not very flexible in the number of basins. Additionally, it requires reasonable settings for the smallest and largest possible niche radii. Note that the clustering is redone after every generation. Hanagandi and Nikolaou [97] incorporate the clustering method employed by Törn [228] into a GA by randomly relocating all but the best individual of every cluster. Gan and Warwick propose the *dynamic niche method* (DNM) [86] as an improvement of [248] in order to provide explicit information about the landscape shape of the treated problem. They employ a variable niche radius and retain niches from the previous generations.

Within his multinational GA, Ursem [236] introduces a purely topological (objective value-based) clustering scheme, the hill-valley method. At first, the whole population resides in one initial cluster. Evaluation of additional search points on a line between two existing ones then reveals where this cluster must be split and a new one emerges: This is the case when the intermediate search point fitness is worse than the linear interpolation of the two line end points. The method is very flexible because the number of clusters is not a priori fixed and no niche radius is needed. However, the main drawback lies in the necessity of additional function evaluations.

It could be discussed whether the *universal evolutionary global optimizer* (UEGO) as suggested by Jelasity [113] is an EA at all, but here we are only interested in its method of detecting basins of attraction. It does so by randomly selecting search point pairs in existing clusters (initially starting with only one) and comparing their fitness values to the one in the middle of their section in the fashion of the hill-valley method (which was to be published one year later, a case of convergent evolution). Additionally, an exponentially decreasing radius scheme is used to merge clusters with overlapping regions. As radii are decreasing in the course of time, smaller and smaller clusters can be established until a user-defined minimal radius is reached.

Gan and Warwick [87] also include the hill-valley method as an additional tool into their DNM basin identification scheme. In [88], they also integrate an agglomerative method termed *niche linkage*, so that clusters of arbitrary shape can result as combinations of already existing clusters. Yao et al. [247] extend the employed hill-valley method in the DNM (or DNC as they term it) with the recursive middling algorithm, which utilizes more internal points between the original search points in order to greatly improve the probability of detecting single peaks. However, this comes at a great cost in function evaluations.

Miller and Shaw [151] suggest *dynamic peak identification* (DPI), which needs a reasonably set niche radius. As in Timmer's MLSL method [226], the initial sample is sorted according to objective values and the available points are put into existing clusters if they are within the niche radius. Shir [206, 207] utilizes this concept within his *ES dynamic niching algorithm*, but combines it with a CMA-ES, thereby establishing a multi-population CMA-ES. The *clearing* scheme of Pétrowski [166], published in the same year as [151], follows a basic principle that is very similar to that of the DPI method, except that it explicitly removes all individuals that are not the best in their niche.

The niching method of the Species Conservation algorithm of Li et al. [125] is very similar to the one of [166], and also needs a pre-specified niching radius. Singh and Deb [212] suggest a *modified clearing*; however, the modification (randomized relocation of cleared solutions) does not apply to basin identification but only attempts to improve the search space coverage and is thus not considered in the following. Li [126] incorporates the species conservation scheme into a differential evolution variant.

Streichert et al. [223] follow a similar idea for obtaining a separation of a sample into niches and apply an agglomerative clustering algorithm, however without respecting the objective values. As in [151] and [166], a niche radius must be provided. However, instead of housing different niches in one huge population, they establish a set of independent subpopulations. This had also been done by Aichholzer et al. [5], but by means of a complete linkage clustering method.

Schaefer, Adamska, and Telega [197] suggest an iterative strategy to recognize basins of attraction one by one with the *clustering genetic search* (CGS). The process has some similarities to the sequential approach of Beasley et al. [26], but the CGS is a niching method as it attempts to reconstruct a full basin map of the search space. Inside the main loop, it runs an EA until enough information is gathered to recognize another basin and then removes its covered search space volume from the actively searched area. This process is repeated until no further basins can be recognized because the remaining landscape is too flat.

The second approach of Pétrowski (together with Genet in this case) [167] uses a classification tree together with a heuristic decision rule, which much better respects the objective values and can thus be put into the group of topology-based methods. However, the proposed method may have problems in search spaces of high dimension as it uses successive hyperplane splits.

Brits, Engelbrecht, and van den Bergh [41] propose a stagnation-based basin identification method for their *niching particle swarm optimization algorithm* (NichePSO). If the objective value of a particle does not change much in some iterations, it is assigned to a new subswarm, together with its closest neighbor. Other particles then may move into the subswarm in the following iterations, and subswarms are merged again if they intersect.

Preuss, Schönemann, and Emmerich [177] suggest the *nearest better clustering* algorithm (NBC) for identifying a variable number of basins by taking into account distances and objective values. The method is extensively discussed in Chapter 4.

The *sample-based crowding* algorithm of Ando, Suzuki and Kobayashi [10] identifies basins with an extended hill-valley method. In order to decide whether two search points reside in the same basin, the fitness integral over the path between them is approximated by sampling additional points near a connecting line. Sample sets of two pairs are compared by employing a Wilcoxon rank-sum test to detect significant differences. When the objective values of one pair are much better than the ones of the other (according to the test), it is said to contain two search points in a basin; these are put into a cluster. The number of samples utilized on each path is relatively high (20 or more), so many additional search point evaluations are needed.

Stoean et al. [219] suggest the *topological species conservation* (TSC) algorithm, which takes up the basic concepts of *genetic chromodynamics* by Dumitrescu [69] and *species conservation* by Li [125] and combines them with the hill-valley method of Ursem [236] in order to identify basins without providing a niche radius. An improved version has been termed TSC2 [221]. Among several modifications, the most important one may be that the number of hill-valley comparisons is kept low by starting with the nearest species first.

Lung and Dumitrescu [135] envision an archive-based strategy termed *roaming* that also combines the hill-valley method with a distance-based criterion for deciding if two search points are located in the same basin.

To remedy the problem of determining a suitable niche radius, Shir et al. [209] introduce a self-adaptive variant where each individual carries (and inherits) its own niche radius. In Shir et al. [210], this concept is refined to allow shape adjustments in order to enable ellipsoid instead of spherical niche shapes. Bird and Li [36] propose a heuristic, density-based scheme for devising a suitable, individual niche radius in their PSO variant. It it based on a near/far notion with the average distances of nearest particle pairs as threshold. If a particle remains near another one for two iterations, the two particles are presumed to be in the same basin of attraction.

On the basis of the *fitness-distance-ratio* (FDR) mechanism of Veeramachaneni et al. [238], which was not intended for niching but for improving convergence speed of a global optimization-oriented PSO, Li [127] introduces a heuristic based on the *fitness Euclidean distance ratio* (FER) value. This is the largest approximate gradient between each particle and its neighbors, so particles are pulled towards their "fittest-and-closest *neighbouring point*". In Zhai and Li [251], the authors show how the ability of PSO niching methods to detect multiple optima quickly can be improved by means of a clever archiving strategy instead of a large population.

Preuss [171] suggests extending the *nearest-better clustering* by a second rule that also takes the number of incoming edges of the nearest-better graph into account, in a manner that reminds one of the *k-topograph* heuristic of Törn and Viitanen [230], but with the much lower $k = 3$.

Qu, Suganthan, and Liang [184] suggest forming neighborhoods (temporary subpopulations to which the variation operators are applied) as in [125], however not by means of a niche radius but from a given fraction of nearest neighbors; besides others algorithms, they propose the *neighborhood based speciation differential evolution* (NSDE) algorithm. In contrast to niche radius-based methods, this does not restrict the area a niche may cover. However, it puts a limitation on the number of basins that may be discovered with a given population size.

Deb and Saha [60] apply a multiobjectivization approach to multimodal optimzation and determine the second, additional objective by counting the better neighbors from each point in the sample. The smaller this number, the more likely it is that the associated point is located near a local optimum. Additionally, they employ a minimal distance parameter to penalize points that are too near to each other. Although Deb and Saha do not explicitly do so, it would be very easy to construct a basin map from the available information.

Epitropakis, Plagianakos, and Vrahatis [76] introduce a nearest neighbor-based concept for their differential evolution variant DE/nrand/1/bin, which is later used as the basic ingredient of the archive-based niching algorithm dADE/nrand/1 by Epitropakis, Li, and Burke [77]. The archive assumes a given niche radius, so the resulting archive contents can be interpreted as a collection of local optima.

Qu, Suganthan, and Das [183] introduce a *locally informed particle swarm* (LIPS) algorithm that replaces the use of the personal best of a particle in the velocity update with averaged local neighborhoods. The resulting algorithm is able to keep the best attained solutions within the population. However, one may argue that in the end, a large set of individuals is attained without any grouping, so there is no explicit basin identification. We therefore do not add the algorithm to Table 5.2.

5.3.4 Comparative Assessment of Niching Method Development

Our niching method overview is far from complete. Even for class A algorithms, we omitted many works that may be considered as niching applications rather than as methodical progress. We also left out methods that focus on detecting funnels, such as the real-valued GA of Oshima et al. [162] or Rönkkönnen's differential evolution variants for multimodal optimization [190] (class C), as they do not perform niching in the sense defined in Sect. 3. Although there is a certain trend in niching methods to employ archives (e.g., Lung et al. [135] and Zhai and Li [251]), we did not enlist any of the works in this direction that would fall into class B, such as Glover's *tabu search* [90], or the *binary space partition trees*-based method of Chow et al. [48].

Note that the latter is not far from meta-modeling-based approaches, which are also omitted here.

One may pose the question of why the niching research area in evolutionary computation is so diverse. The two main reasons we see have already been stated above and are again considered here from a more abstract perspective:

- Many discoveries have been made several times in different communities. There was not simply not a single niching research area but there were many, and little interaction existed between them. Looking at Table 5.2, we find that topology-based techniques (employing search space distances as well as objective values) and graph-based techniques known in global optimization since the 1980s and 1990s, respectively, have become more and more important within several branches of evolutionary computation.

- For a long time it was, and to a certain extent still is, unclear what the exact task of niching methods shall be. With the use cases defined above we have provided an overview of the different requirements for the solutions obtained by niching methods. Any new works will clearly define what their new algorithms are aiming for. Our view is that while transferring mechanisms found in natural evolution to evolutionary computation is important for advancing EC methods, mimicking nature is not a valuable task in itself in the context of optimization, although several works argue differently, such as Della Cioppa et al. [61]. Biology is surely a valuable source of inspiration but our methods have to work in a mathematical environment, and being closer to nature does not reliably lead to improvements in optimization performance. We also regard all requirements towards *how* to do niching as obstructive. In a variation of Mahfoud's [136] niching target statement, Li [128] states that niching methods should be "*able to locate multiple optima and maintain these found optima until the end of a run*". We disagree. For the end user, it is irrelevant when and how the desired multiple optima have been detected and where they have been kept as long as they are available at the end of the optimization process.

Our impression is that after some years of "*mutating and recombining*" niching methods, we have now reached a consolidation phase. Performance comparison of niching methods as fostered by the CEC 2013 niching methods competition of Li, Engelbrecht, and Epitropakis [129] will be important for recognizing the most promising paths for further research. We believe that whatever new algorithms come up in the near future, they have to take into account the topology of the treated optimization problems, namely the relation of distances and objective values.

Table 5.2 Niching methods employed in evolutionary computation and global optimization, classes A, B, C according to Sect. 5.3. If available, we refer to the name of the niching mechanism if applicable, and the algorithm otherwise. *Dist.* and *obj.* stand for the use of distance and objective value information, respectively; an "*i*" means that this information is only used indirectly, usually for sorting. *K var* means that the number of basins to detect is not fixed a priori but determined by the niching method. The *basic technique* gives the underlying technique; for class A approaches, this usually refers to some cluster analysis method.

year	method name	class	dist.	obj.	k var	basic technique
1970	alg. of Becker and Lago [27]	A	✓	i	✓	density based clustering
1973	Törn's LC algorithm [228]	A	✓	i	✓	density based clustering
1975	crowding [56]	C				local selection
1984	single linkage GOA [226]	A	✓	i	✓	single linkage clustering
1984	multi-level single linkage [226]	A	✓	✓	✓	topological & single-link
1987	sharing [91]	C				selection modification
1992	topographical GO [230]	A	✓	✓	✓	topological
1993	sequential niching [26]	B			✓	derating
1993	adaptive clustering [248]	A	✓			k-means
1994	tagging [215]	C			✓	randomized
1996	dynamic peak identificat. [151]	A	✓	i	✓	single-link
1996	clearing [166]	A	✓	i	✓	single-link
1998	UEGO [113]	A	✓	✓	✓	topological & single-link
1998	SGA-CL [97]	A	✓	i	✓	density based/Törn LC
1999	hill-valley method [236]	A		✓	✓	topological
1999	shifting balance GA [161]	B	✓		✓	island location control
1999	classificat. tree speciation [167]	A	✓	✓	✓	topological
1999	dynamic niche method [86]	A	✓	i	✓	topological
2000	$\kappa(\mu(\tau)/\rho,\lambda)$-ES [5]	A	✓		✓	complete linkage
2001	DNM wt. hill-valley [87]	A	✓	✓	✓	topological
2002	NichePSO [41]	A	✓	✓	✓	stagnation & single-link
2002	DNM/niche linkage [88]	A	✓	✓	✓	topological & single-link
2002	species conservation [125]	A	✓	i	✓	single-link
2003	clustering based niching [223]	A	✓		✓	single-link
2004	clustered genetic search [197]	A	✓	i	✓	density based clustering
2005	ES dynamic niching [206]	A	✓	i		single-link
2005	nearest-better clustering [177]	A	✓	✓	✓	topological
2005	sample-based crowding [10]	A		✓	✓	topological
2005	DE species conservation [126]	A	✓	i	✓	single-link
2006	DNM wt. recursive middl. [247]	A	✓	✓	✓	topological
2006	ES adaptive niching [209]	A	✓	i		adaptive single-link
2006	adaptive niching PSO [36]	A	✓			adaptive single-link
2007	fitness-Euclidean dist.ratio [127]	A	✓	✓	✓	topological
2007	roaming [135]	A	✓	✓	✓	topological & single-link
2007	topological species cons. [219]	A		✓	✓	topological
2010	ES shape adaptive niching [210]	A	✓	i		adaptive single-link
2010	topological species cons. 2 [221]	A	i	✓	✓	topological
2011	dynamic archive [251]	A	✓	✓	✓	adapt. s.-link/stagnation
2011	NOAH [235]	C	✓			density based removal
2012	nearest-better clustering 2 [177]	A	✓	✓	✓	topological
2012	neighborhood based SC [184]	A	✓	i	✓	single-link
2012	multiobjectivization [60]	A	✓	✓	✓	topological
2013	dADE/nrand/1 [77]	A	✓	✓	✓	adaptive single-link

Chapter 6
Nearest-Better-Based Niching

Here we employ the nearest-better clustering basin identification method derived in a previous chapter for setting up two niching evolutionary algorithms. After doing parameter testing, we investigate how these algorithms perform in comparison to other recent methods for the all-global and one-global use cases by means of available benchmark suites.

6.1 Two Niching Methods and Their Parameters

How can the parallel and sequential approaches be realized in NBC-based niching algorithms, and how should their parameters be set?

If we envision different ways to employ the *nearest-better clustering* (NBC) method that was investigated in Chapter 4 as a niching method within an optimization algorithm for multimodal problems, two major concepts come to mind:

- We can attempt a parallel search that tries to obtain many optima concurrently, as demanded by Mahfoud [136]. An algorithm that follows this scheme was suggested and experimentally assessed in Preuss [170] (there termed NBC-CMA-ES).

- Alternatively, we may use NBC to detect good starting points for sequential runs of a local optimization procedure as envisioned with the *niching evolutionary algorithm 2* (NEA2) in Preuss [171]. The layout of this algorithm deliberately disregards Mahfoud's demand and NEA2 therefore delivers local optima one by one.

In the following, we will derive two different niching algorithms that realize these two concepts, and then discuss some possible extensions and reasonable parameter settings.

139

Applying NBC to a sample of points in the search space delivers a clustering that hopefully corresponds well to the basin structure of the optimized problem. Regardless of whether these clusters are treated in parallel or sequentially, we have to choose a local search procedure for detecting the local optima. From the many available possibilities, we employ only the *covariance matrix adaptation evolution strategy* (CMA-ES) for this purpose, for the following reasons:

- The CMA-ES is considered the strongest available evolutionary black-box optimization algorithm, and we explicitly target black-box problems, with no derivative information available. We presume that this assessment has also led to employing the CMA-ES within the related niching methods of Shir et al. [209, 210], which strengthens the argument.

- Although we presume that the tackled problem is moderately multimodal, we cannot safely assume that there is really only one local optimum located within the area that is covered by one cluster (see also the error rates obtained in Experiment 4.6.4.1). For this reason, a method that still has some global search capabilities and is not instantly stuck if an area with almost zero gradient is reached will be favorable.

- The stop/restart conditions of the CMA-ES are very well developed. Especially if we want to find many optima in a sequential style, this is very important, because all function evaluations wasted here delay the next searches.

We utilize the CMA-ES in the form laid out in Auger and Hansen [13], but without increasing the population size. The only other change we make is that it gets the inhabitants of a detected cluster as the start population, and not a single point. The population size is constantly kept at the default $\lambda = 4 + \lfloor 3 \ln D \rfloor$, $\mu = \lfloor \frac{\lambda}{2} \rfloor$. Here, λ denotes the number of offspring individuals generated in one generation, and μ stands for the number of parent individuals. D refers to the number of dimensions of the treated problem. If the number of cluster points is smaller than μ, we simply start with a smaller population and the CMA-ES completes it in the next generation. If it is larger, we select the μ best search points and discard the rest. Of course, we disable restarts within the CMA-ES, but the restart conditions come in handy to stop the search; we simply assume that the CMA-ES has reached the desired local optimum.

6.1.1 Niching Evolutionary Algorithm 1

This method is designed to work in parallel in multiple basins of attraction according to the categorization given in Sect. 1.3.4. It largely complies with the NBC-CMA-ES algorithm that was introduced and experimentally investigated on the BBOB problem set in Preuss [170]. The only important difference is that we have exchanged the NBC clustering method with the newer variant that employs *rule 2*,[1] given in Algorithm 3 in Chap. 4 (instead of Algorithm 2 in Chap. 4).

Algorithm 5: NEA1

---

1 distribute an evenly spread sample over the search space, evaluate it;
2 **repeat**
3   apply NBC: separate sample into populations according to clusters;
4   **forall the** *populations* **do**
5   perform one CMA-ES generation (variation/selection);
6   mark population inactive if any CMA-ES stop criterion true;
7   collect all surviving individuals from the active populations;
8 **until** *no active populations left*;
   // start all over:
9 **if** *!termination* **then**
10   goto step 1

Algorithm 5 shows the single steps of NEA1 in pseudocode. We start with a rather large sample (on the order of $50D$) spread over the whole search space. Then, a loop is entered where the nearest-better clustering is applied to the available general population, which is the whole sample in the first iteration and the set of surviving individuals of all active populations after selection in all subsequent iterations. From the obtained clustering, separate populations are built, one for each cluster. In practice, it has been shown to be advantageous if the number of populations is bounded, because otherwise, we may get so many of them that it is not possible to get near to the sought optima within the available budget of function evaluations. However, this depends on the environmental conditions (optimized problem, needed accuracy, budget) and usually only occurs for highly multimodal functions. As a rule of thumb, one may set a limit of 100 concurrent populations. If there are already existing populations (from the second iteration on), an attempt is made to conserve these as much as possible.

We then enter a second loop, where each population is run within a separate CMA-ES for only one generation. If at least one of the restart criteria for a population is true after the selection step, the state of this population is set to *passive* and it is disregarded in the following steps. After passing through all populations, the selected individuals are collected in a general population and if there is at least one active population left, we start the outer loop again by applying the nearest-better clustering.

If the available budget for function evaluations is large enough, we may have a situation where no active populations are left although we did a continuous redistribution of individuals to populations in every iteration. In that case, the algorithm does a complete restart and starts with a new initial sample. It may occur advantageous to keep some of the attained information and do a more informed restart, but a lot

---

[1] Rule 2 takes the indegree of nodes in the nearest-better graph into account.

Algorithm 6: NEA2

---

1 distribute an evenly spread sample over the search space;
2 apply NBC: separate sample into populations according to clusters;
3 **forall the** *populations* **do**
4 | run a CMA-ES until any CMA-ES stop criterion is hit;

   `// start all over:`
5 **if** *!termination* **then**
6 | goto step 1

---

of testing with different variants of archives has revealed that (besides making the algorithm much more complex) this rarely has a positive effect.

6.1.2 Niching Evolutionary Algorithm 2

The NEA2 algorithm (see Algorithm 6) was suggested in Preuss [171] as an improvement over the NEA1 algorithm. As we will see in the following, it indeed performs better under most conditions. In addition, it immediately becomes clear that it has an even simpler structure than NEA1. The major difference between the two algorithms is that in NEA2, we attempt to search sequentially and not in parallel. The approach has the advantage that the first local optimum is usually detected much earlier than for NEA1. This is especially useful if only very short response times can be tolerated or the algorithm is run on a unimodal problem.

After starting with a large sample over the whole search space as in NEA1, we again apply NBC to generate separate populations that should match the basin structure of the tackled problem as well as possible. These populations are then used sequentially as initial populations for separate runs of a CMA-ES, which are continued until a stop criterion is met. When all populations have been run and there are still function evaluations left, the whole process is started again with step 1. As with NEA1, attempts to take over information from the first iteration when doing a complete restart have not shown consistent positive effects, and are therefore disregarded here.

6.1.3 Parameter Settings and Extensions

Fortunately, we have already fixed most of the parameter settings for the NBC itself, as documented in Figure 4.14 and Figure 4.16. These rules depend only on the number of dimensions of the treated problem (which is known) and the population size. For the local optimization procedure (CMA-ES), it is known that the population size and the initial step size are the most important parameters. Supporting experimental results for this assumption can be found in Preuss [169], Preuss, Rudolph, and Wessing [176], and Wessing, Preuss, and Rudolph [240], besides others.

As stated above, the population size employed by the CMA-ES itself shall be kept at its (rather small) dimension-dependent default value. Small CMA-ES populations enable us to run many local searches. Additionally, the main reason for using large population sizes within the CMA-ES is to enhance the global optimization capabilities, e.g., for coping with highly multimodal functions. However, in our case, these difficulties shall be treated by the clustering method, and choosing large populations for the local search will only slow down the detection of local optima. But how shall we set the size of the initial sample that provides the data for the first clustering? As the NBC parameters depend on this number, its interactions with the algorithms may be complex. We decide to subject the NEA1 and NEA2 algorithms to a parameter tuning process in order to detect the initial sample size, the initial step size, and two more parameters that are explained further down.

For both algorithms, we restrict the number of clusters that are recognized to 100. All additional clusters are simply ignored, which means that the corresponding search points are not carried over to the next generation. However, NEA1 rarely hits this limit, and NEA2 almost never does.

6.1.3.1　Using an Archive

When thinking of possible extensions of our niching methods, it may be tempting to integrate an archive that could be used for coordinating restarts. One could also imagine that adding old search points (such as the initial samples of the global restarts) may improve the obtained clustering, but unfortunately, this is not true in general. Our experiments have shown that there are situations in which the algorithm can benefit from an archive, but these are not encountered very frequently. Besides, the search space volume is growing so rapidly for more than two dimensions that many evaluated search points are needed to obtain a good covering. This may be the reason why archives are so rarely used in niching, as documented in Table 5.2. The *roaming* method of Lung and Dumitrescu [135] and the *dynamic archive* for PSO of Zhai and Li [251] are the only niching methods that make use of an archive for controlling the search process.

Archives may provide some potential for improving niching algorithms in the future, but it seems that more research is needed to find out how to use them efficiently. We therefore use an archive only for collecting the obtained local optima. These are recorded whenever a restart criterion of the CMA-ES is met.

6.1.3.2 Adapting the Step Size to the Estimated Basin Size

If we assume that the obtained clustering is accurate enough to allow for estimating the sizes of the corresponding basins, we can try to use this information to control the mutation step sizes of the single CMA-ES runs employed for local optimization. The rationale behind this idea is that we want to prevent visiting already known areas (other basins) again and therefore limit the step sizes.

For NEA2, this concept was first realized in Preuss [171], by simply limiting the initial step size of each CMA-ES run to the average distance of the cluster individuals from their center. A related idea was suggested in Loshchilov et al. [131], namely reducing the initial step size σ^0 to smaller values when the population size is increased after a restart (for the NIPOP-CMA-ES) or sampling the initial step size within an interval that is limited from above by the default step size (for the NBIPOP-aCMA-ES).

For NEA1, the clustering is not regarded as stable as for NEA2, as it is redone in every generation. Therefore, we do not employ the current population diameters but only the average distance to the nearest recorded local optimum that is found in the archive. Of course, this means that in the first generations, where the archive is still empty, the step size control does not apply.

For both algorithms, we now introduce a parameter (named *sigmaToDistance*) that is used for logarithmic scaling of the computed maximal step size σ_{max}, such that $\sigma_0 = \sigma_{max} * 10^{sigmaToDistance}$. This parameter is also used in the following NEA1/NEA2 parameter tuning investigation. Besides possibly detecting a better relative upper bound for the step size, we can also find out if limiting the step size makes sense at all by varying the *sigmaToDistance* parameter. If our initial reasoning is correct, the parameter will provide the best algorithm performance at values around zero. Much larger values would mean that the limitation itself is counterproductive, much smaller values that the step size must be set in an even more restrictive fashion.

6.1.3.3 Using Additional Evaluations for Basin Detection

Within the experiments presented in [178], we have seen that the NBC clustering algorithm matches well with the additional use of the *detect-multimodal* (DMM) mechanism[2] in order to improve the correspondence between basins and clusters. The detect-multimodal (DMM) mechanism takes two evaluated points in the search space and tests the one that lies exactly between them. Four result states are possible:

1. The intermediate point is better than both original ones, so all presumably follow the same peak.

---

[2] This is another name for the hill-valley detection method of Ursem [236].

2. The intermediate point is worse than both original ones; thus these almost surely follow different peaks (an error may be made here if a single basin of attraction possesses a very complicated geometrical shape, but even in this case it may be reasonable to regard it as two basins for the optimization process).

3. If the intermediate point is between the two original ones in quality but worse than the linear interpolation, it is presumably located at the outer area of a basin (where the gradient is getting smaller beyond a turning point; this employs the assumption of an approximate Gaussian shape of the basin).

4. If the intermediate point is between the two original ones in quality but better or equal than the linear interpolation, it is presumably located in the inner area of a basin.

Note that the two latter cases are usually not treated separately from the first one in existing algorithms. Consequently, if a type 1, 3, or 4 point is detected, we also add the newly evaluated additional search point to the cluster the first DMM input point is put into. However, testing with the DMM method needs great care; otherwise many function evaluations may be wasted. We therefore apply the test only to the direct neighbors in the sorted list of edges that is available in step 5 of the NBC algorithm (algorithm 2 in Chap. 4), and likewise for the NBC algorithm with rule 2. Only the edges that would otherwise be cut are tested, and if the DMM test signals that the cut is unnecessary as the two points probably belong to the same basin, the cut is reversed.

This needs some more evaluations but provides greater confidence in the quality of the obtained clustering because it can detect type 2 (clusters cover more than one basin; also see Sect. 4.6.3) and type 3 errors (multiple clusters in one basin), so a clustering generated by NBC can be corrected. Attempting to employ DMM only (without NBC preprocessing) would often lead to a similar result but waste a large number of evaluations (see [178]) and is thus not realized here. Note that in order to keep the number of additional evaluations as low as possible, we always evaluate only one intermediate point. For the same reason, we also dismiss the recursive middling algorithm by Yao et al. [247]. It would be more accurate than the original DMM, but require many more evaluations.

For the following parameter experimentation, we envision a probabilistic variant of using the type (1–4) information. The parameter *multimodalCutProbs* may have values from $[0, 3]$, and two search points are regarded as belonging to different clusters under the following conditions (let r be a random number from $[0, 1]$):

- If *multimodalCutProbs* is smaller than 1 and the DMM result state is either not 1 or r is larger than *multimodalCutProbs*, the cluster is cut into two.

- If *multimodalCutProbs* ≥ 1 and < 2, we cut if DMM $\neq 1$ and (DMM $\neq 3$ or $r \geq$ (*multimodalCutProbs* -1)).

- If *multimodalCutProbs* ≥ 2, we cut if DMM $= 2$ or (DMM $\neq 4$ or $r \geq$ (*multimodalCutProbs* -2)).

If none of the conditions apply, the cut that would otherwise have been done is reversed as the two points presumably reside in the same basin. The mapping to a real-valued range from 0 to 3 may appear somewhat complicated, but enables the tuning method to test several ways of using DMM information and the transitions between them at once. We seamlessly blend completely ignoring the DMM result state (for *multimodalCutProbs* = 0), applying the DMM in the standard way (for *multimodalCutProbs* = 1), and taking into account DMM result states 3 and 4. Note that a similar probabilistic parameter mapping has also been employed in Chimani, Kandyba, and Preuss [46].

6.1.3.4 Experiment 6.1: Find a Robust Default Parameter Setting for NEA1/NEA2

Pre-experimental Planning

In preliminary parameter scans based on *Latin hypercube designs* (LHDs), the necessary run length for obtaining the global optimum on the considered problems (the ones from Experiment 4.6.4.1 plus the BBOB functions f15-Rastrigin rotated, f20-Schwefel's $x\sin(x)$, and f21-Gallagher's 101 Gaussian peaks) was determined to be $3 \cdot 10^5$, with a required precision of 10^{-4} (this suffices for reaching the global optimum in most cases). The BBOB functions are chosen in order to represent very different scenarios that may be encountered when optimizing multimodal problems. Note that for the BBOB functions, the required precision is usually set to 10^{-8}, but we reduce it deliberately here as we are more interested in the ability of the algorithms to detect the global optimum, not to approximate it with very high precision. We also assume that the parameter settings obtained for the *one-global* case will be reasonable defaults for the other cases as we strive to avoid bad or too specialized parameter settings rather than do fine tuning.

With some SPO test runs, a reasonable setting of the SPO parameters was determined. Note that our approach is brute force rather than fast tuning in order to achieve a reliable result; we therefore choose a the relatively high number of ten initial repeats.

Table 6.1 SPO parameter range for the NEA1/NEA2 default parameter search

parameter	min	max	comment
EAinitSigmas	0.1	2.0	initial step size for CMA-ES runs
initialPopsize	10	80	initial sample size, multiplied by D
sigmaToDistance	-2.0	2.0	step size to basin size adaptation
multimodalCutProbs	0.0	3.0	use of detect-multimodal (DMM) information

Task

For each algorithm, we strive for a robust parameter setting that should if possible work well for 2D and 5D, for all six treated problems. Rather than choosing high-performance parameter combinations, we avoid parameter ranges that lead to bad performance in at least one case. In the case of conflicting parameter suggestions, we choose values closer to the center of the parameter interval.

Setup

All single NEA1/NEA2 runs are executed until $3 \cdot 10^5$ function evaluations are used up or the global optimum has been attained with precision 10^{-4}. The expected running time (Equation 5.1) for reaching this target is employed as a performance criterion. The ranges of the four parameters to be tuned are given in Table 6.1. SPO is run as described in Bartz-Beielstein and Preuss [24], with ten initial repeats on an LHD of size 500, and only one new suggested parameter configuration per iteration. Every single parameter combination may only be run up to a maximum of 50 repeats (SPO increases the number of repeats for the most successful parameter settings) and SPO is allowed a budget of 10,000 single NEA1/NEA2 runs.

The tuning is done for NEA1 and NEA2 in 2D and 5D for all six problems (see above) separately, and the parameter value choice with the necessary aggregation of results for different functions is done later on in a manual fashion based on the guidelines given in the task description. The search space is bounded to $[-5,5]^D$ for the BBOB problems and to $[0,20]^D$ for the MPM/FMPM problems; thus its size is largely similar. It was seen in Mersmann et al. [147] that 2D is a good representative of 3D algorithm behavior as well, whereas 5D is more similar to 10D and 20D in terms of problem hardness; therefore we restrict the tuning to these two settings.

Results/Visualization

The results of the tuning processes are displayed in Figure 6.1 (2D) and Figure 6.2 (5D) for NEA1 and in Figure 6.3 (2D) and Figure 6.4 (5D) for NEA2. Only parameter configurations that have been repeatedly tested by SPO are considered in order to concentrate on the more successful variants; thus a large part of the initial LHD is omitted.

The box-percentile plots show each single parameter effect in five rows that represent equally sized groups of parameter configurations, sorted according to their performance. Each dot stands for a single parameter configuration, the surrounding structure gives the percentiles with the vertical bar approximately positioned at the middle, representing the median value, and the large dot represents the mean value.

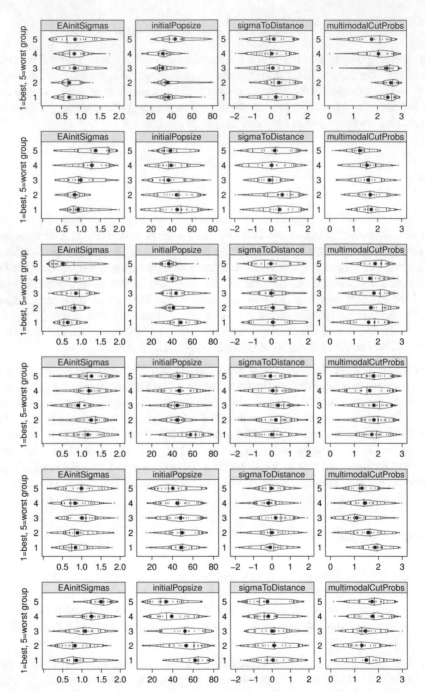

Fig. 6.1 Box percentile plots for the four parameters for NEA1 in 2D on test problems (top to bottom): f15 (Rastrigin rotated), f20 (Schwefel's $x\sin(x)$), f21 (Gallagher's Gaussian 101 peaks), all from the BBOB benchmark suite, and FMPM 5/100, MPM 20, and MPM 100.

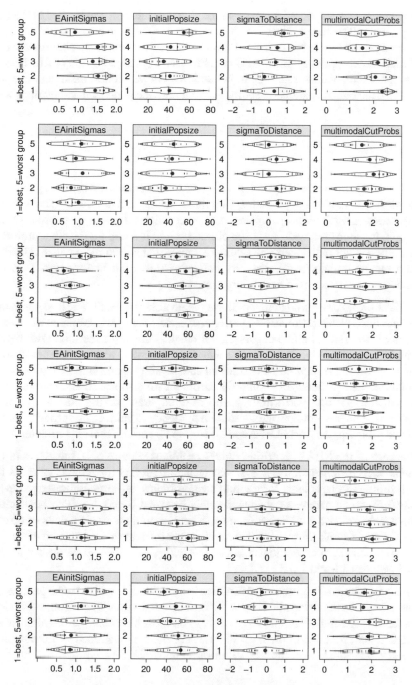

Fig. 6.2 Box percentile plots for the four parameters for NEA1 in 5D on test problems (top to bottom): f15 (Rastrigin rotated), f20 (Schwefel's $x\sin(x)$), f21 (Gallagher's Gaussian 101 peaks), all from the BBOB benchmark suite, and FMPM 5/100, MPM 20, and MPM 100.

The parameter values should be chosen such that they are well represented in the best groups, over all problems. Note that the columns of the whole box-percentile plot table represent the four tested parameters, while the rows stand for the six problems, so that each row corresponds to one SPO run. For each algorithm it is of specific interest to compare the obtained rankings in 2D and 5D. As we need to make a consensus decision (rule out bad parameter settings), we ignore the parameter interactions here.

Observations

For the NEA1 tuning on 2D problems (Figure 6.1), we find that the best parameter settings vary a lot over the problems. The most emphasized difference can be found for the *multimodalCutProbs* parameter, where very high values are appropriate for the Rastrigin function but a wide spectrum of good values is found for the other problems. Generally, a small initial step size seems recommendable, and the mean and median values for the initial sample size vary between 40 and 60. The *sigmaToDistance* parameter does not seem to have a large impact.

For NEA1 on 5D problems (Figure 6.2), we obtain a similar result. The best parameter configurations also have relatively high *multimodalCutProbs* parameter values and the initial step sizes shall be chosen rather small. The only exception with regard to the 2D results can be found for the Rastrigin function, where higher step sizes seem to be favorable in 5D. The best initial sample sizes are even more concentrated around the interval between 40 and 60, although we also see some more extreme values within the best performing group of configurations (the bottom row with number 1). The best configurations show a tendency to values larger than 0 for the first two problems and to values around 0 for the *sigmaToDistance* parameter.

In the case of the NEA2 algorithm on 2D problems (Figure 6.3), the effect of the *multimodalCutProbs* parameter is much weaker than it is for NEA1, whereas the good configurations at least for the first two problems have much more condensed *sigmaToDistance* parameter values, namely between 0 and 1.5. The best initial sample sizes are also a bit larger than for NEA1, leaning towards values of 60 and more for the last four problems and a much larger interval centred around 40 for the first two problems. For the initial step size (*EAinitSigmas*), the best values are mostly found between 1.0 and 1.5.

On the 5D problems (Figure 6.4), NEA2 seems to be less sensitive to the parameters in most cases, as the percentile shapes for the best group (row 5) are generally wider than for 2D. The initial sample size seems to be most important, with values of 60 favored at least for three of the six problems. For the other parameters, the values of the best group spread over most of the allowed interval.

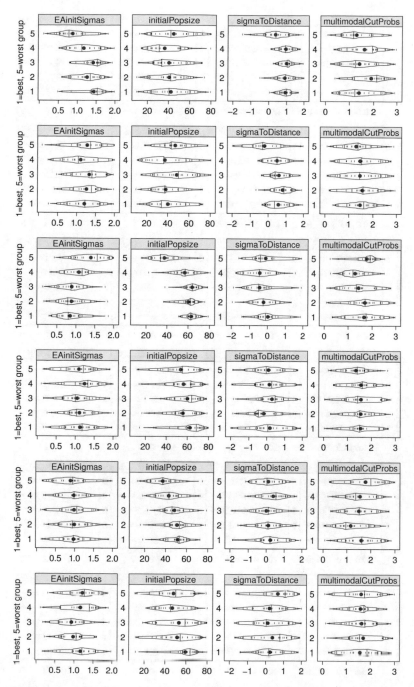

Fig. 6.3 Box percentile plots for the four parameters for NEA2 in 2D on test problems (top to bottom): f15 (Rastrigin rotated), f20 (Schwefel's $x\sin(x)$), f21 (Gallagher's Gaussian 101 peaks), all from the BBOB benchmark suite, and FMPM 5/100, MPM 20, and MPM 100.

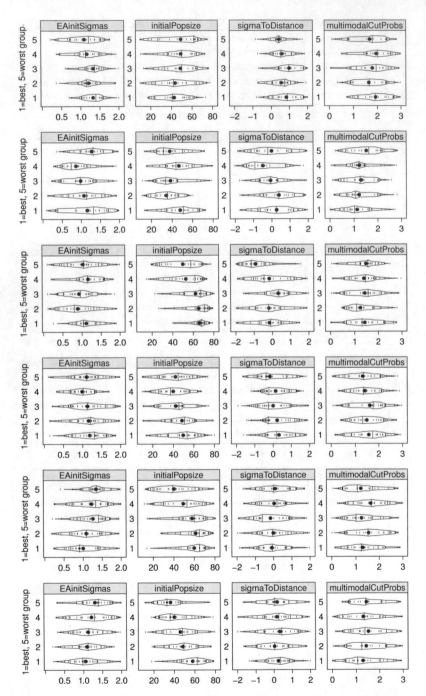

Fig. 6.4 Box percentile plots for the four parameters for NEA2 in 5D on test problems (top to bottom): f15 (Rastrigin rotated), f20 (Schwefel's $x\sin(x)$), f21 (Gallagher's Gaussian 101 peaks), all from the BBOB benchmark suite, and FMPM 5/100, MPM 20, and MPM 100.

Discussion

From the obtained results, it immediately becomes clear that it is not possible to derive a one-fits-all parameter setting for NEA1 or NEA2 that reliably achieves the best possible performance even for only these six test problems. It seems that whenever the tuning process is run on a single problem, we mostly obtain different best parameter configurations. However, we can of course give some recommendations for a reasonably well-performing setting that should work in many cases. We may state that too extreme settings are suitable only for some parameters for some of the test problems, and employing parameter values located in the middle of the tried intervals appears to be much safer if problem-specific settings are not known. Note that for each of the problems, we know that the single parameter values of each row belong together, but we did not investigate which of the dots exactly belongs to which other in the plot for another parameter. Thus, we can basically reason only based on the parameter *intervals* of the best or second best groups. Nevertheless, this simplification enables us to generalize over the test problems, which we basically do by detecting the intersections of these intervals.

For the initial step sizes, we find a slight tendency for larger values for NEA2 than for NEA1. This may be due to the fact that the number of clusters obtained within a run is generally higher for NEA1, as it applies the clustering method after each generation, whereas an established clustering is never refined in NEA2. We therefore recommend an initial step size of around 1.0 for NEA1 and 1.25 for NEA2. These are absolute values and most likely depend on the search space size and thus shall be scaled for much smaller or larger spaces.

The obtained results show that the initial sample (population) size should be somewhere between $40D$ and $60D$ for most test problems; thus we set it to $50D$ for both algorithms. For some problems, larger values seem appropriate. However, a larger setting would probably waste a lot of evaluations for other problems, especially if the budget for available evaluations is small. For very short runs, one may still reduce the sample size to $40D$ or even smaller values. As the initial sample must be large enough for the NBC to work, it should not be smaller than around 50 search points.

For the *sigmaToDistance* parameter, it is interesting to see that the idea generally works, although it seems that the exact setting does not have a major impact on most test problems. If the accomplished limitation of step sizes based on estimated basin sizes were to largely fail, the best parameter configurations would exhibit values of at least around 1.0 or larger, thereby disabling the limitation. This is obviously not the case. As the parameter values for the best groups are well spread for most of the test problems, we stay with the original default of 0.0 for this parameter.

The *multimodalCutProbs* parameter is the only case of a qualitative difference in successful settings between NEA1 and NEA2. This may be due to the fact that the underlying DMM mechanism is used with different data in both algorithms. For NEA1, we perceive a tendency to larger values in the best group from the visualizations, whereas for NEA2 we have an interval spread over nearly all possible values in all cases but Rastrigin $5D$. For this reason, we decide to switch off the DMM mechanism completely for NEA2 (set this parameter to 0.0), and to use a default parameter value of 2.0 for NEA1.

The recommended parameter settings are summarized in Figure 6.5 and also used in the following experiments, unless otherwise stated.

6.2 Performance Assessment for the *One-Global* Case

How do the NEA1 and NEA2 algorithms perform on the BBOB problem set?

Since the first issue of the *Black Box Optimization Benchmarking* (BBOB) workshop at the GECCO 2009 conference, its 24 functions set and experimental setup (the 2009 functions and setup are described in Hansen et al. [102, 100]) has been established as a quasi-standard for measuring the global optimization performance of evolutionary and related algorithms. In the following years, 2010, 2012, and 2013, the BBOB workshop was run with only very slight changes. The 2013 setup can be found in Hansen et al. [101], and it features an additional second set of post-processing tools for the expensive optimization case (up to $100D$ function evaluations). The *Comparing Continuous Optimisers* (COCO) webpage[3] provides a lot of data, including the results of all algorithms ever submitted to a BBOB workshop.

As the function set has been kept unchanged, it is possible to compare new algorithms to the already recorded data. The functions themselves are implemented as generators, so a large number of slightly different function instances can be employed. Over the years, only the instance set has been changed (in 2010, instances 1 to 15, in 2012, instances 1 to 5 and 21 to 30), but results on different instance sets are generally regarded as similar enough to allow for comparisons. The strengths of the BBOB set are that, besides its setting a standard, it prevents a number of setup mistakes

---

NEA1: initial step size 1.0
      (based on bounded spaces with edge length of 10 to 20 per dimension),
      initial sample size $50D$
      sigmaToDistance parameter 0 (previous default)
      multimodalCutProbs parameter 2.0

NEA2: initial step size 1.25
      (based on bounded spaces with edge length of 10 to 20 per dimension),
      initial sample size $50D$
      sigmaToDistance parameter 0 (previous default)
      multimodalCutProbs parameter 0.0 (DMM switched off)

Fig. 6.5 Recommended default parameter values for NEA1 and NEA2

---

[3] http://coco.gforge.inria.fr/

observed in earlier experimental studies on evolutionary algorithms. The global optima are never 0.0, and their optimizers are never found at the origin. Additionally, the measuring is not done in the optimization algorithms but transferred to the problem code. Post-processing and visualization are also done by the BBOB code, so that a number of issues that could result from mistakes in measuring are circumvented. In comparison to older experimental studies, a lot more data is taken into account; namely, the number of evaluations to reach 50 objective value targets is recorded, and not just a single number such as the objective value obtained after a number of defined function evaluations.

6.2.1 Choosing a Set of Multimodal BBOB Test Problems

From the viewpoint of niching methods, which are not especially targeted at providing only one very good solution, the BBOB test set has several disadvantages. To begin with, only the global optimization performance is measured, so that obtaining several good solutions is not taken into account at all. Then, the 24 employed functions contain a lot of separable and/or unimodal problems and very specific cases that are interesting for analyzing algorithm performance under extreme conditions (such as problems with high conditioning). Of the five problem groups, only the last two are interesting when measuring niching methods. These multimodal problems are divided into one group with "*adequate global structure*" and one with "*weak global structure*".

Group 4 of the BBOB test set consists of the following functions (note that we take over the original numbering and thus start with function 15):

15 Rastrigin function
16 Weierstrass function
17 Schaffer's F7 function
18 Schaffer's F7 function, moderately ill-conditioned
19 Composite Griewank-Rosenbrock function F8F2

Group 5 of the BBOB test set contains these test problems:

20 Schwefel function $\sum x_i sin(\sqrt{|x_i|})$
21 Gallagher's Gaussian 101-me peaks function
22 Gallagher's Gaussian 21-hi peaks function[4]
23 Katsuura function
24 Lunacek bi-Rastrigin function

---

[4] Note that *me* and *hi* stand for medium and high conditioning around the global optimum, for function 21 the conditioning is about 30, for function 22 it is about 1000.

A global structure is usually understood as a recognizable shape in the set of local optimizers, the simplest of which could be a parabola, as with the Rastrigin function. It was shown in Mersmann, Preuss, and Trautmann [148] that the existence of a global structure is an important property of an optimization problem that can influence the performance of algorithms strongly. Thus, it makes sense to divide the multimodal problems into these two groups. However, the property differences are not as clear as one may think. Whereas the global structures for functions 15 and 17 to 19 are unimodal, this is not the case for the Weierstrass function, which is repetitive (contains multiple global optimizers). Function 23 is also highly repetitive with more than 10^D global optimizers, but the global structure of function 24 is only bi-modal and the one of Gallagher's Gaussian functions is completely random. For function 20, the best peaks are located at opposing sides of the search space, so one may speak of a deceptive global structure, reminding us a bit of a Rastrigin function turned upside-down.

Of the ten test problems in the two multimodal groups, only four are really interesting for the performance analysis of niching algorithms, namely functions 16, 20, 21, and 22. As we have learned from the results of the NBC-CMA-ES on the BBOB functions in Preuss [170] and from tests with early NEA1/NEA2 variants in Preuss [171], our algorithms do not perform well on highly multimodal problems with a simple global structure, such as functions 15, 17, and 18, and it seems that function 24 (Lunacek bi-Rastrigin) is similar enough to 15 to be very difficult for niching methods. The key to solving function 23 is using very small initial step sizes. In addition to being confirmed by our own preliminary experiments, this is also confirmed in Loshchilov et al. [131]. Most likely, this stems from the abundant number of local optima that enforce very small step sizes because otherwise the algorithm would due to the repetitive structure of the problem be continuously jumping around between different areas that look similar. Interestingly, detecting the global optimum of function 16 (Weierstrass) seems to be easier although the two share some similarities. We presume that the main difference is that there are considerably fewer optima in function 16 and therefore the basins are larger.

6.2.2 Measuring the One-Global Performance

Unfortunately, no basin or optima location information is available for the BBOB implementation, so it is not plausible to use the problem set for any other than the *one-global* case. One cannot even apply a measure that uses no problem knowledge, e.g., the R5S measure that was suggested in Sect. 5.2.4, because no archive of points is recorded for each run, but only the subsequent improvements are.

6.2.2.1 Experiment 6.2: On Which of the BBOB Functions Does Niching Speed Up Finding Only the Global Optimum?

Pre-experimental Planning

The studies in Preuss [170, 171] serve as preparatory experiments in this case. Based on these results, we fix the maximum number of evaluations to 10^6. The test functions are determined based on the reasoning above. One may object that the test problems are chosen such that good results can be expected. However, we have simply selected the functions that represent the type of problems our methods have been designed for.

Task

We demand that at least for one of the four selected test functions, our niching methods (specifically NEA2, which should be faster if only the global optimum is pursued) perform clearly better than the corresponding IPOP-CMA-ES and BIPOP-CMA-ES algorithms, as these use almost the same local search method, except with slightly different initial conditions (random location, with adaptation of the population size).

Setup

The NEA1 and NEA2 algorithms are run on the BBOB functions 16, 20, 21, and 22 with the parameters specified in Figure 6.5 for up to 10^6 evaluations (each run is stopped earlier if the global optimum has been attained), with an accuracy of 10^{-8}. For means of comparison, we use the data of the IPOP-CMA-ES, BIPOP-CMA-ES, and GLOBAL algorithms from the COCO repository. The latter method is described in detail in [164] and goes back to the algorithm of Boender [37], which is also contained in Rinnooy Kan, Boender and Timmer [189]. It employs single linkage clustering in the global phase and a quasi-Newton method with a *Broyden-Fletcher-Goldfarb-Shannon* (BFGS) update for the local searches. As is usual for the BBOB setup, the performance is measured by the *expected running time* (ERT, Equation 5.1) until the global optimum is reached.

Results/Visualization

Although each algorithm has only been run once (on 15 different instances per problem), we provide the ERT up to the global optimum being reached with accuracies

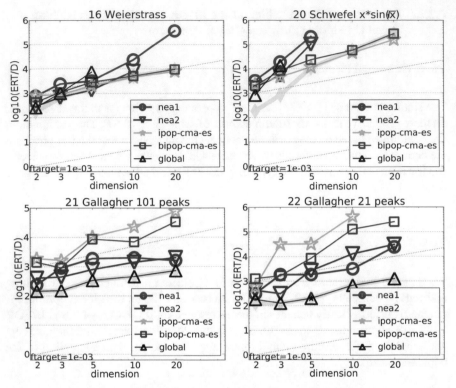

Fig. 6.6 $\log 10(ERT/D)$ values of the five compared algorithms on the BBOB functions 16, 20, 21, and 22, up to a target value of 10^{-3} plus the global optimum. Dimension D on the x axis. Missing entries mean that the corresponding method has not attained the global optimum reliably in 15 runs (possibly due to a too restrictive choice of run length). The light brown lines connect the best reached values (marked by the diamonds) for any algorithm of the GECCO 2009 workshop, separately determined for each target value and number of dimensions.

10^{-3} and 10^{-8} in Figures 6.6 and 6.7, respectively. The accuracy information was not given to the algorithms, which could otherwise have stopped searching earlier by applying an accuracy-based restart criterion. The figures display $\log 10(ERT/D)$ over the dimensions $D \in \{2, 3, 5, 10, 20\}$.

Observations

The first thing to note is that the groups for the two target values in Figures 6.6 and 6.7 very much resemble each other. For functions 16 and 20, there are only slight differences. Up to five dimensions, the Weierstrass function is reliably solved by all algorithms, with only marginal performance differences. In ten and 20 dimensions, the CMA-ES variants take the lead, and the GLOBAL method obviously does not

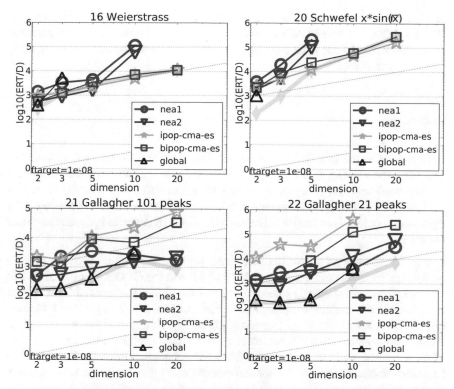

Fig. 6.7 $\log 10(ERT/D)$ values of the five compared algorithms on the BBOB functions 16, 20, 21, and 22, up to a target value of 10^{-8} plus the global optimum. Dimension D on the x axis. Missing entries mean that the corresponding method has not attained the global optimum reliably in 15 runs (possibly due to a too restrictive choice of run length). The light brown lines connect the best reached values (marked by the diamonds) for any algorithm of the GECCO 2009 workshop, separately determined for each target value and number of dimensions.

reach the target value within its allowed budget (of $2 \cdot 10^4 D$). NEA1 and NEA2 both perform similarly, with a small advantage to NEA2 up to $10D$. For function 20, we obtain a similar picture, but here the CMA-ES variants already perform much better from $5D$ on.

On the Gallagher functions with their relatively small number of randomly located peaks, GLOBAL is the best of the five algorithms and only fails for higher precision in $20D$. For the 101 peaks problem (function 21), NEA2 is faster than NEA1 in $3D$ and $5D$ and both perform similarly in the other cases. In $10D$, the two algorithms are comparable to the GLOBAL algorithms when a smaller distance (10^{-8}) from the global optimum is required. With the exception of $3D$, NEA1 and NEA2 are both faster than the two CMA-ES variants. On function 22 (21 peaks), all algorithms are very close together in $2D$ for the lower precision. For the higher precision, the IPOP-CMA-ES performs remarkably badly. In $3D$, NEA2 is a bit faster than NEA1 and the BIPOP-CMA-ES for both precisions, and from $5D$ on, NEA1 and NEA2 perform consistently better than the BIPOP-CMA-ES, with a small advantage to NEA1 in 10D. The IPOP-CMA-ES always stays clearly behind all other algorithms, which is most likely due to its growing population that ineffectively uses up too many function evaluations.

Discussion

With respect to the task defined above, we can state that for two of our four test problems, the niching algorithms perform better than the two canonical CMA-ES variants with very few exceptions. Interestingly, the difference is much clearer for the lower precision (10^{-3}) target values. It seems that the niching methods detect the right basins quickly but then need more evaluations to descend towards the global optimum.

Comparing the results on the 101 peaks and the 21 peaks functions shows that the performance differences between NEA1/NEA2 and the GLOBAL method shrink, and that NEA2 gets better in relation to NEA1 for higher numbers of local optima. It seems that the niching methods inherit the local search ability of the CMA-ES and combine it with the clustering-based global search capability of the GLOBAL method. We conclude that this combination is the more useful the higher the number of local optima and the higher the demanded accuracy. However, the results of functions 16 and 20 show that if the number of basins gets too large, niching has at least no advantage and fails in high-dimensional spaces (from around 5D on).

Even taking modern CMA-ES variants such as the NIPOP-CMA-ES or the NBIPOP-aCMA-ES from the BBOB 2012 results into account, our niching methods perform remarkably well on the Gallagher functions. The two new variants employ the negative update of the covariance matrix as introduced by Jastrebski and Arnold [112] and further developed in Hansen and Ros [104]. Introducing these novel CMA-ES extensions would most likely also lead to improved speed of the local searches within NEA1/NEA2. However, we did not attempt this because we want to investigate what the consequences of applying a niching method are, and this is documented well by comparing to the older IPOP-CMA-ES and BIPOP-CMA-ES data.

The CMA-ES as local search component within the NEA1/NEA2 algorithm provides a lot of flexibility concerning the properties of the fitness landscapes, as the CMA-ES copes well with a certain amount of ruggedness and noise, and also adjusts to extreme differences in scaling between different variables. However, it could be advantageous to exchange this component or at least complement it with a quasi-Newton or gradient-based method. The resulting method would surely be faster as long as the basin shapes are smooth, but lose ground as soon as one of the difficulties named above occurs.

6.3 Performance Assessment for the *All-Global* Case

How do our algorithms perform on the problem set of the 2013 IEEE CEC competition on Niching Methods for Multimodal Optimization?

Whereas the status of black-box optimization benchmarking is commendable, the same cannot be said about the comparison of niching algorithms in evolutionary computation. Despite an abundant number of publications, there are few attempts to enable a fair and standardized comparison of niching methods. Most researchers simply collect their own set of functions and use their own experimental setups, so most of the available data is not comparable. The 2010 IEEE *Congress on Evolutionary Computation* (CEC) featured a special session on niching methods for multimodal optimization, but the first organized competition took place only in conjunction with CEC 2013.[5]

6.3.1 The CEC 2013 Niching Competition Problems

For experimentally investigating the performance of our two algorithms in situations when more than one good solution needs to be obtained, we employ the problem set and setup of Li, Engelbrecht, and Epitropakis [129]. They utilize 12 different functions that are gathered from various sources, from $1D$ up to $20D$. Optimization algorithms are expected to deliver a result set that contains the best obtained points (currently, the size of this set is not penalized) and write it to a file. The niching competition software then computes the number of optima contained in each result set, for five levels of accuracy $\varepsilon \in \{10^{-1}, 10^{-2}, \ldots, 10^{-5}\}$. For each level, the peak ratio is computed as the average number of obtained peaks divided by the number of global optimizers, over 50 repeats. The competition rules state that the overall winning algorithm should be the one with the highest average peak ratio over all problems and accuracies. In the following, we will be more interested in the performance of different algorithms on the specific functions and thus look at the single peak ratios or the ranking of algorithms on single functions. However, we are very grateful to Michael Epitropakis and Xiaodong Li of the competition organizational team for providing some preliminary data we can compare the results of our algorithms to.

One may argue that the competition setup has several weaknesses, at least if compared to the BBOB setup. The major deficiencies are the lack of function instances and the missing bias in the locations of the global optimizers and also in function values. Adding a bias has the advantage that it is impossible for the optimization algorithms to "guess" where the global optimizers are and what function values they have. We are also not completely satisfied with the performance criterion. Instead of the peak ratio, the *peak distance* (see Sect. 5.2.2) could be used, and very good local

---

[5] http://goanna.cs.rmit.edu.au/ xiaodong/cec13-niching/competition

optima could also be taken into account. Additionally, the time issue is currently completely ignored (when are the optimizers found?). However, the competition setup is currently the only reasonable, publicly available benchmark suite for niching algorithms, and it may evolve and improve in the next years.

Table 6.2 CEC 2013 niching competition problem set with additional numbering for referral in result figures. *Dim* stands for the number of dimensions, *global opt.* the number of global optimizers, and *max evals* the allowed number of fitness evaluations.

ID function	name	dim	global opt.	max evals
1 F_1	five-uneven-peak trap	1D	2	5.0E4
2 F_2	equal maxima	1D	5	5.0E4
3 F_3	uneven decreasing maxima	1D	1	5.0E4
4 F_4	Himmelblau	2D	4	5.0E4
5 F_5	six-hump camel back	2D	2	5.0E4
6 F_6	Shubert	2D	18	2.0E5
7 F_6	Shubert	3D	81	4.0E5
8 F_7	Vincent	2D	36	2.0E5
9 F_7	Vincent	3D	216	4.0E5
10 F_8	modified Rastrigin - all global optima	2D	12	2.0E5
11 F_9	composition function 1	2D	6	2.0E5
12 F_{10}	composition function 2	2D	8	2.0E5
13 F_{11}	composition function 3	2D	6	2.0E5
14 F_{11}	composition function 3	3D	6	4.0E5
15 F_{11}	composition function 3	5D	6	4.0E5
16 F_{11}	composition function 3	10D	6	4.0E5
17 F_{12}	composition function 4	3D	8	4.0E5
18 F_{12}	composition function 4	5D	8	4.0E5
19 F_{12}	composition function 4	10D	8	4.0E5
20 F_{12}	composition function 4	20D	8	4.0E5

6.3.1.1 Experiment 6.3: How Do NEA1/NEA2 Perform in Comparison to Their Non-niching CMA-ES Counterparts and Other Recent Methods for Multimodal Optimization?

Pre-experimental Planning

During our first experiments, we found that although especially NEA2 performs well on most test functions, it makes sense to increase the tolerance (*TolFun*) value for detecting stagnation in the CMA-ES local searches. The standard parametrization would be TolFun $= 10^{-12}$, as suggested in Hansen [98]. As the required precision for the CEC niching benchmark is never below 10^{-5}, we set this value to TolFun $= 10^{-6}$. This means that a local CMA-ES run is stopped if all observed function values for the last $10 + \lceil 30D/\lambda \rceil$ generations and the current generation have a range of less than 10^{-6}. The parameter setting applies to all our methods that employ the CMA-

ES as a local search mechanism: the default CMA-ES, the IPOP-CMA-ES, NEA1, and NEA2. Preliminary results show a performance improvement (in average peak ratio) of around 2%, compared to the original setting. We assume that this change is acceptable even from a real-world application perspective, as the user is asked only for the minimal distance in objective values that is still of interest.

Task

With this experiment, we mainly want to answer two questions. The first is: are NEA1/NEA2 better than their non-niching CMA-ES counterparts in finding multiple optima at once? Note that this is an *experimentum crucis* for the niching hypothesis: If our algorithms fail, we have no experimental evidence to support the meaningfulness of spatially distributed restarts in comparison to randomly placed ones. If they are successful, we have at least identified situations in which the niching hypothesis holds. In order to test the niching hypothesis, we require that NEA1 and NEA2 be significantly better (as indicated by a Wilcoxon rank sum test with significance level of 5%) than the random restart CMA-ES in at least 75% of the cases. This level may appear arbitrary, and in fact there is no strong motivation to choose exactly this number, but we want to know if there is a substantial difference. So the criterion shall be somewhere between 50% and 100%. For reasons of simplicity, we choose the center of this interval.

The second question is: how do our methods perform compared to other recent methods for multimodal optimization? We will look at each of the five accuracy levels 10^{-1} to 10^{-5} separately and determine a ranking of all available algorithms for each problem, and then add up the ranks for each accuracy level. The task for NEA1/NEA2 is to achieve the lowest rank sums over all accuracies.

Setup

All available algorithms (NEA1, NEA2, CMA-ES, IPOP-CMA-ES, and the *topological species conservation 2* (TSC2) from Stoean et al. [221]) are run on the 20 problems with 50 repeats for the number of function evaluations specified in Table 6.2. TSC2 is a niching EA that heavily employs the DMM/hill-valley mechanism of Ursem to keep the population distributed in many basins throughout an optimization run; also see Sect. 5.3.3. By CMA-ES we simply mean an IPOP-CMA-ES without increasing population size.

For NEA1 and NEA2, all results of local search processes (whenever a population is stopped due to CMA-ES restart conditions) are recorded in an archive, and this archive is provided to the niching competition software for detecting the obtained optima. As for the BBOB experiment, they use the parameters specified in Figure 6.5

with the above-mentioned stop condition parameter $\mathtt{TolFun} = 10^{-6}$. The CMA-ES and the IPOP-CMA-ES are run with the same archiving and stop condition parameters ($\mathtt{TolFun}$ set to 10^{-6}); otherwise they use their own default values as given in Hansen [98]. The only difference in the default parameters for all four algorithms is in the initial step size, which is chosen relative to the search space size as $\sigma_{init} = 0.2(b_{upper} - b_{lower})\sqrt{D}$, with b_{upper}/b_{lower} standing for the upper and lower search space bounds, respectively, and D meaning the number of dimensions. This roughly corresponds to the step sizes employed for the BBOB experiment and specified in Figure 6.5. Note that such a parameter setting in the dependency of the search space bounds would not be possible if the bounds were not known.

The parameters for all other methods are collected in Table 6.3. For TSC2, we choose a parameter set that emerged as successful in the parameter investigations of Stoean et al. [221]. The performance data of all *differential evolution* (DE) methods on the CEC 2013 niching competition problem set has generously been provided by Michael Epitropakis. Of these, three algorithms have been invented by Rönkkönen [190] and are nicely summarized in Rönkkönen [191]:

DECG: crowding DE with gradient descent

DELG: local selection differential evolution with gradient descent

DELS: local selection differential evolution

These variants change the more globally oriented character of DE to a more locally oriented one and partly hybridize it with a simplified gradient descent method that executes only a single line search in the direction of the approximated gradient whenever the local phase is selected.

DE/nrand/1/bin and DE/nrand/2/bin are suggested by Epitropakis, Plagianakos, and Vrahatis [76] as descendants of the classic DE/rand/1/bin and DE/rand/2/bin variants of Storn and Price [222], and the basic difference is that they choose the nearest

Table 6.3 Parameter settings for TSC2 and DE-based methods: *pc*, *pm*, *ms*, and *grad* mean recombination probability, mutation probability, mutation strength, and gradations, respectively. The DE parameter F stands for the mutation or scaling factor, and *CR* for the probability of generating an offspring individual instead of keeping the parent.

algorithm	parameter settings
TSC2	population size 150, pc=0.4, pm=0.6, ms=0.5, grad=2
DECG	population size 100, F=0.5, CR=0.9
DELG	population size 100, F=0.5, CR=0.9
DELS/adaptive jitter	population size 100, F=0.5, CR=0.9
DE/nrand/1/bin	population size 100, F=0.5, CR=0.9
DE/nrand/2/bin	population size 100, F=0.5, CR=0.9
dADE/nrand/1	population size 100, F, CR adapted as in the JADE [252] algorithm
dADE/nrand/2	population size 100, F, CR adapted as in the JADE [252] algorithm

neighbor of a current individual instead of a random individual as the base vector when computing new trial vectors (offspring).[6]

By adding the dynamic archive of Zhai and Li [251] to the DE/nrand/1/bin algorithm, Epitropakis, Li, and Burke [77] obtain dADE/nrand/1. The archive employs a niche radius R, and new solutions can enter the archive if either they are better than stored solutions in their vicinity (given by R), or they represent previously unexplored regions. The related dADE/nrand/2 algorithm by Epitropakis is currently unpublished but similar to dADE/nrand/1, except that it uses the DE/nrand/2/bin scheme as the base method instead of DE/nrand/1/bin. In our view, the dynamic archive makes these methods true niching methods. The ancestor methods given above would not fall into this category.

We follow the competition manual [129] in selecting the peak ratio as performance criterion, recorded only after the complete number of function evaluations per function. For NEA1, NEA2, CMA-ES, IPOP-CMA-ES, and TSC2, we thus obtain the single peak ratios for every run, whereas for the DE variants, we only have the aggregated data (average peak ratios). In order to answer the two above-mentioned questions, we separate the following sections into a phase 1 (concerning our own results) and a phase 2 (taking into account the DE results).

Results/Visualization Phase 1

Fig. 6.8 Overview of the pairwise rank sum tests between the algorithms, performed separately on each of the 20 problems and five accuracy levels. Of the 100 cases, we remove the ones for which both algorithms reach all optima in all repeats (this mainly applies to the very simple first five test problems. Then, the fraction of the significantly better with regard to the remaining cases is computed in both directions. Each row displays the ratio of significantly worse cases, each column the significantly better cases in comparison to all other algorithms. NEA2 is significantly better than all other algorithms in at least 60% of the cases.

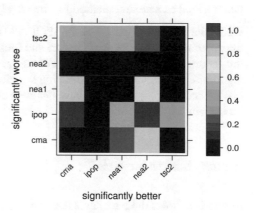

---

[6] Note that the DE naming scheme is derived from the operators used within a DE method, the number stands for the number of difference vectors generated for each trial individual, and *bin* means binary tournament selection.

Figure 6.8 provides the aggregated pairwise Wilcoxon rank-sum tests between NEA1, NEA2, CMA-ES, IPOP-CMA-ES, and TSC2.

Observations Phase 1

The best algorithms are NEA2 and the CMA-ES, where in the direct encounter NEA2 is significantly better than the CMA-ES in around 60% of the counted cases. NEA1 performs worse than these two, but it is still better than the IPOP-CMA-ES and TSC2 in around 90% of the counted cases.

Discussion Phase 1

NEA2 slightly misses the task of being significantly better than the CMA-ES in at least 75% of all cases; it achieves a bit more than 60%. However, as the CMA-ES is only better in less than 10% of the cases, the niching algorithm can safely be considered superior under this setup (in the remaining cases, the rank-sum test does not indicate significant differences). Concerning NEA1, we find a much worse ratio: It is significantly better than the CMA-ES in about 20% and worse in about 60% of the cases. We presume that the relatively small number of pursued global optimizers for most problems and the relative tight budget of function evaluations prevent NEA1 from reaching its full potential. However, due to its sequential nature, NEA2 combines the best of both worlds (niching and fast local search) and is therefore better suited as a default method for the *all-global* case.

While the bad performance of the IPOP-CMA-ES is easy to explain as it doubles the population size after every restart and thus uses up the available budget quickly, the failure of TSC2 is much more suprising. As we will see in the following, the algorithm has problems approximating the global optimizers closely enough for the hardest accuracy levels, which have not been employed in the original publication of Stoean et al. [221].

Results/Visualization Phase 2

Figure 6.9 shows the rank sums over all algorithms and problems, computed separately per accuracy level. Figure 6.10 displays the peak rates of all algorithms on all problems for the easiest and hardest accuracy levels.

Fig. 6.9 Rank sums of the average peak ratios of all 12 algorithms over all 20 problems, separately computed for each accuracy level (lower is better). In the case of ties due to equal peak rates (as often happens for the first five problems), all algorithms get the average value of the ranks occupied by the tied entries. The best achievable rank sum is thus 15, the worst 240. For accuracy level 10^{-1}, dADE/nrand/1 (dade1 in the figure) achieves the lowest rank sum of 69, followed by NEA2 with 76, and dADE/nrand/2 with 79.5. For all other accuracy levels, NEA2 performs best with rank sums between 60.5 and 67.

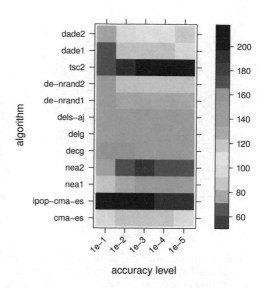

Observations Phase 2

We can make two interesting observations in Figure 6.9. First, average performance seemingly divides the algorithms into three groups. The best group consists of NEA2, the two dADE/nrand variants, and the CMA-ES. TSC2 and the IPOP-CMA-ES perform worst, and all other algorithms form the middle group (green). Secondly, several algorithms show a trend toward getting worse in rank if the accuracy levels get harder, which means that in absolute numbers, they get worse faster than the other algorithms do. This behavior appears to be especially strong for the dADE/nrand variants, NEA1, TSC2, and the IPOP-CMA-ES.

Figure 6.10 explains the last observation in more detail: Whereas NEA2 and the CMA-ES reach almost the same peak ratios for both accuracy levels (and thus probably also for the intermediate ones), the dADE/nrand variants perform worse on most problems (except for the first five) for the hardest accuracy level, as also NEA1 and TSC2 do. Interestingly, none of the 12 algorithms obtains any global optimizer for the hardest accuracy level on problem 6, which is the *2D* Shubert function with only 18 global optimizers.

Discussion Phase 2

If we consider that the dADE/nrand algorithms are basically DE/nrand variants with an added archive functionality, the performance shift obtained by this modification is remarkable. The dADE/nrand/1 method clearly dominates all other algorithms with

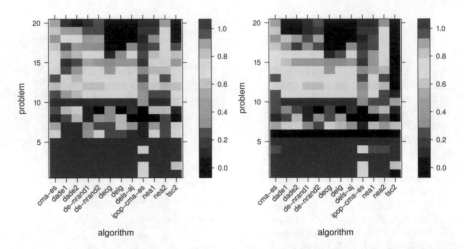

Fig. 6.10 Peak rates of the 12 compared algorithms on the 20 functions of the CEC 2013 niching competition. Left: required accuracy 10^{-1}; right: 10^{-5} (number of fitness evaluations fixed). A value of 1.0 means that all peaks have been approximated with the required accuracy, a value of 0.0 (black) that no peak has been found. Note that the number of target peaks is different as indicated in Table 6.2.

parallel search character (the other DE-based methods, NEA1, and TSC2). Figure 6.9 shows that for the first accuracy level, it is the best of the 12 algorithms, closely followed by NEA2. However, it obviously converges slower to the optimizers than the sequential methods (NEA1 and the CMA-ES). For the relative low number of global optimizers sought in this experiment, the sequential approach seems to be advantageous, and (as already stated in phase 1), niching does increase the chance to find several good solutions more quickly.

Concerning the original task for NEA1/NEA2 defined above, namely achieving the lowest rank sums over all accuracies, this is clearly missed for NEA1, but almost reached by NEA2, which appears to be the algorithm of choice for a large range of problems and accuracy levels.

Note that the CMA-ES parameters have not been tuned for the problems treated here. Doing so would probably benefit also NEA1 and NEA2 as they employ the CMA-ES as a local search method. It may well be that such tuning would have led to different best parameter settings for NEA1/2 and the two CMA-ES variants. Especially for the latter, we would expect an emphasis on small populations. However, tuning all parameters of all methods would be beyond the scope of this experiment and we would be surprised to see large differences in the algorithm ranking resulting from it.

6.4 Conclusions

In this chapter, we first introduced our own niching algorithms NEA1 and NEA2, which operate on the basis of the *nearest-better clustering* (NBC) method derived in Sect. 4. The basic difference between the two approaches is that NEA1 operates in a *breadth-first search* (BFS) fashion, and NEA2 realizes a *depth-first search* (DFS). Extensive parameter testing was performed on problems which are generally thought of as being suitable for niching algorithms, integrating several ideas for extending the two methods. The most successful of these ideas may be the limitation of step sizes to the estimated basin size, an approach that is currently also being pursued by other researchers. Whereas integrating Ursem's hill-valley/detect-multimodal mechanism is of limited use for NEA1 and seemingly unsuitable for NEA2, the use of archives may be an interesting path to follow in the near future. While in the performance comparisons on a subset of the BBOB problems (*one-global* case) we did not use any archive, a very simplistic archive was employed for the CEC 2013 niching competition experiment (*all-global* case). However, the example of the dADE/nrand/1 algorithm of Epitropakis, Li, and Burke [77] shows that adding a clever archiving strategy can strongly improve the performance of a formerly mediocre algorithm.

Unfortunately, there are currently no benchmark suites for the *all-known* and *good-subset* cases, and we deliberately abstain from setting these up in this book. We see such an endeavor rather as a community effort than as the output of a single researcher. Additionally, it would not be too difficult to change the CEC 2013 niching competition setup of Li, Engelbrecht, and Epitropakis [129] to also cover these two cases. The main difficulty that has to be resolved for getting there is agreeing on suitable performance measures. The work of Preuss and Wessing [180] establishes a good overview of available measures and could be helpful in this respect. As real-world applications often require a certain number of alternative good solutions instead of a single best one, we see a lot of potential in continuing efforts to improve and compare niching algorithms.

The experimental results of this chapter have shown that NEA2 performs at least as well as NEA1 on the BBOB test problems, and much better on the CEC 2013 niching test problems. This advantage stems from the sequential approach that pursues every local search process to stagnation instead of evolving a whole population until most of the interesting solutions have been reached concurrently. The only major weakness of the suggested niching methods is their inability to deal with a strong global structure in a highly rugged landscape, such as that of the Rastrigin function.

Summarizing, we claim that for obtaining a reasonably small set of very good solutions on a not-too-multimodal problem of restricted dimensionality (up to about $20D$), employing NEA2 can be highly recommended. As it basically combines a random sample with clustering and subsequent single CMA-ES runs, it is easy to understand and implement (CMA-ES code is available for many programming languages), and it can directly profit from further CMA-ES developments. Alternatively, another local search mechanism could be used inside NEA2, so that it may also be applied to non-real problem representations. Due to its sequential nature, NEA2 nearly performs as well as the CMA-ES on unimodal problems, which is a desirable side effect.

Chapter 7
Summary and Final Remarks

With the materials provided in the six previous chapters, we hope to have extended and deepened knowledge about evolutionary algorithms applied to multimodal black-box optimization problems. It is now time to summarize the main insights, review what we achieved in comparison to the original goals listed in Sect. 1, and give some hints concerning promising future search directions. Before we turn to the last of these three issues, we will deal with the first (summary) in the context of the second (goals), as each of the chapters is tightly linked to one of the goals. The only exception to this is Sect. 2, which presents our view on the current state of an experimental methodology for research in evolutionary computation and related fields. It may be seen as a summary of the many experiences with experimental research the author has been involved with in the last few years. As the amount of available theory concerning evolutionary computation applied to multimodal problems is small, it is quite important to do experiments in a structured and well-defined way and we make an effort to respect this throughout the whole work.

7.1 Goal 1: Improve the Understanding of Niching in Evolutionary Algorithms and Evaluate Its Potential Benefits

After presenting the foundations of black box optimization and also evolutionary computation and its roots, we provided a classification of (evolutionary) methods for multimodal optimization based on five types of parallelization in Sect. 1.3.4. This suggested classification scheme is to our knowledge original and more complete than any other that may have been suggested before, and the main types of evolutionary algorithms that are applied to multimodal problems are represented.

After summarizing the current biological concepts of niching and speciation and their counterparts in evolutionary computation in Sect. 3, we introduced the notions of basin identification and basin recognition and set up a simple niching model that enables us to compare the efficiency of parallel and sequential randomized searches. When considering heuristic optimization algorithms on multimodal problems, it is important to keep the outlined four phases of the optimization process in mind: searching for the global optimum (t_2) requires that we cover just half of the available basins on average, whereas guaranteeing its detection (t_3) requires that we visit all basins, which is a lot harder. Based on theoretical approaches for the coupon collector's problem (CCP), t_2 and t_3 can be computed directly for simple cases. For more complex cases, e.g., unequal basin sizes, we have to rely on experimentation. As an overall conclusion from that chapter, it is becoming clear that niching is of limited use if only one global optimizer is sought. However, as soon as a good coverage of the existing basins of attraction is desired in order to deliver multiple good solutions, niching should be helpful.

Although this niching model was suggested (albeit in a much simpler form) in [168], we are not aware of any attempts of others to develop it further. However, even the extensions presented in this work do not enable us to answer all relevant questions and improving the model could be rewarding for obtaining a better understanding of the underlying processes of niching algorithms.

The performance measure and algorithm overview given in Sect. 5 shall help us actually shape the niching research field. We define four different scenarios concerning the task of the optimization process, and three different classes of techniques dealing with multimodal problems, of which only the first one could be called niching algorithms by our understanding. Interestingly, there have been several older algorithms from the field of global optimization (partly dating back to the 1970s) that fulfill our criteria for niching algorithms, whereas some newer algorithms from the evolutionary computation field that are explicitly called niching algorithms do not. It becomes clear that while we attempt to provide the possible alternatives for performance measures and also a taxonomy of the niching algorithms, this can be only a first step towards consolidation or unification, which is, by our understanding, what the field really needs. The CEC 2013 niching optimization competition[1] was a good step in this direction, as researchers were forced to apply the same measures on the same problems to make the results comparable. However, the competition could still be improved in many ways, and we hope this will happen in the next few years.

7.2 Goal 2: Investigate Whether Niching Techniques Are Suitable as Diagnostic Tools

In Sect. 4, we presented the nearest-better clustering (NBC), which was invented in [177] and refined in [171]. We added a correction mechanism for large samples to keep the influence of artifacts of the random distribution employed for determining

---

[1] http://goanna.cs.rmit.edu.au/ xiaodong/cec13-niching/competition

the initial population small. On a number of test problems, the parameters for the NBC algorithm were determined with respect to two criteria, namely the useful cluster fraction and the detected basin fraction. It turns out that with a well-chosen parameter setting, a good approximation of the number of basins can be obtained, but only for lower-dimensional problems ($5D$ still works well, $10D$ is already very difficult). While we provided experimentally determined, reasonable default values for the NBC, the capability of the method to reveal problem properties was not the most important aspect of the analysis. We are sure that the NBC can be made better use of in this respect, but we cannot follow this path further within this work. As an interesting future development, we suggest using the clustering byproducts in the context of the exploratory landscape analysis (ELA) approach, as they come at no additional evaluation effort. Getting an estimation of the modality of a problem would surely help when choosing a suitable optimization algorithm.

7.3   Goal 3: Compare the Performance of Niching and Canonical Evolutionary Algorithms

This goal was pursued in Sect. 6, where our two niching-based optimization algorithms (NEA1 and NEA2) were presented and experimentally evaluated in a global optimization context (BBOB), and on the problems of the CEC 2013 niching competition. Meanwhile, NEA2 officially won this competition, performing better than NEA1, the different CMA-ES and DE variants described in Sect. 6.3, and some other algorithms submitted to the competition. From all these results, it becomes clear that our niching methods are very suitable for scenarios in which several good solutions are needed. In the global optimization case (BBOB), they perform well only if the optimized problems match their design assumptions well: the problem should have a moderate degree of modality and not be high-dimensional ($D \leq 20$). If these conditions hold, NEA2 usually performs better than NEA1 and also better than the random restart CMA-ES variants it itself uses as local search methods. Note that recently, several refinements of the CMA-ES have been proposed and also measured on the BBOB test set. However, exchanging our CMA-ES code with that of a newer variant within NEA2 and then comparing the performance of this CMA-ES variant without any niching should not change the results much, as the improvements mainly affect the local search performance. In order to improve the optimization performance of NEA2 even further, this would indeed be an interesting direction to follow in the future.

7.4 Goal 4: Estimate for Which Problem Types Niching EAs Actually Outperform Canonical EAs

The model-based investigations of Sect. 3 and the performance tests with full niching-based optimization algorithms in Sect. 6 are well in agreement concerning the situations where niching EAs can be expected to be better than canonical EAs. Of the four scenarios defined in Sect. 5, we have solid evidence for at least two (also see the discussion for goal 3):

- For the one-global scenario, niching methods are only advantageous if the problem is not too multimodal and not too high-dimensional. This scenario corresponds to time t_2 in the above-mentioned model. Of course, niching methods will not perform well on unimodal problems.

- In the all-global case, niching algorithms can be expected to perform very well, usually better than other EAs or other global optimization methods. This scenario corresponds to time t_3 for which the model also predicts a much larger potential for improvements. However, for practical implementations of niching algorithms, there will be a dimension limit of around 20. Beyond that limit, it gets very difficult to do any sort of basin identification, so the niching methods will most likely fail (or revert to the behavior of a randomly restarted local search technique if any such method is employed).

Interestingly, the all-global scenario does not differ much from the all-known case in the model, which leads us to conjecture that niching algorithms will often be advantageous here, under the same conditions as for the all-global case. We expect a similar result also for the good-subset scenario, but also foresee a much higher dependency on the chosen performance measure. It will be rewarding to explore these last two cases more in the coming years.

7.5 Conclusions

During the course of this work, several algorithmic ideas were tried and did not work. While the overall goal always stayed the same—namely obtaining a good EA for multimodal problems—our preferred algorithm evolved considerably. NEA2 is the method that finally came out of this process, and it is surprisingly simple. Many complex variants or auxiliary mechanisms that were tried have perished due to weaker performance. Under the conditions discussed above ($D \leq 20$, multimodal with medium number of optima), we recommend using NEA2 or a similar niching-based algorithm. The advantage of these methods will be even more pronounced if several very good solutions are desired, not only one.

However, the expected improvement depends on many conditions, of which the most prominent may be the performance criterion. Further experimentation is needed to achieve more reliable predictions. Competitions such as the CEC 2013 niching optimization competition will be a good way to explore the interplay between algorithms and performance measures in the future.

Another important issue is the application of NEA2 to real-world problems. As it only needs a distance metric to run the NBC clustering method in order to detect the basins of attraction, it may well be applied also to mixed-integer or even structural optimization problems, given that the local search method (CMA-ES for real-valued search spaces) is replaced by one that matches the representation. The example of Ulrich's NOAH method [235] shows that algorithms that at least sustain diversity can come up with surprisingly different, good solutions.

If we follow the literature overview given in Sect. 5, it is obvious that there seems to be very little theory on niching methods (at least for real-valued search spaces, the situation changes a bit if combinatorial/discrete search spaces are considered). The older global optimization papers from the previous century contain at least some interesting theoretical approaches, but it seems that the EC community never took these up. However, it would certainly be nice to have at least some basic theory so that not every new question requires setting up a new experiment, which is not at all a trivial endeavor, as we hope became clear in Sect. 2.

Last but not least, after the collaboration with Ofer Shir on [211] and Simon Wessing on [180], we are more than ever convinced that multimodal and multiobjective optimization have many things in common and should learn from each other. In both cases, a population with a certain search space spread is sought, except that the exact conditions of goodness are different. We expect that exchanging measures and methods between these fields will lead to considerable progress in the years to come.

References

1. B. Addis. *Global Optimization using Local Searches*. PhD thesis, Department Systems and Computer Science, of University of Florence, 2005.
2. B. Addis and M. Locatelli. A new class of test functions for global optimization. *Journal of Global Optimization*, 38:479–501, 2007.
3. B. Addis, M. Locatelli, and F. Schoen. Local optima smoothing for global optimization. *Optimization Methods and Software*, 20(4-5):417–437, 2005.
4. C. C. Aggarwal, A. Hinneburg, and D. A. Keim. On the surprising behavior of distance metrics in high dimensional spaces. In *ICDT '01: Proceedings of the 8th International Conference on Database Theory*, pages 420–434. Springer, 2001.
5. O. Aichholzer, F. Aurenhammer, B. Brandtstätter, T. Ebner, H. Krasser, and C. Magele. Niching evolution strategy with cluster algorithms. In 9th *Biennial IEEE Conf. Electromagnetic Field Computations*, 2000.
6. E. Alba and B. Dorronsoro. *Cellular Genetic Algorithms*, volume 42 of *Operations Research/Computer Science Interfaces*. Springer, 2008.
7. E. Alba and M. Tomassini. Parallelism and evolutionary algorithms. *IEEE Transactions on Evolutionary Computation*, 6(5):443–462, 2002.
8. M. Ali and C. Storey. Topographical multilevel single linkage. *Journal of Global Optimization*, 5(4):349–358, 1994.
9. R. Anderson. The role of experiment in the theory of algorithms. In *Proceedings of the 5th DIMACS Challenge Workshop*, volume 59 of *DIMACS: Series in Discrete Mathematics and Theoretical Computer Science*, pages 191–196. American Mathematical Society, 1997.
10. S. Ando, E. Suzuki, and S. Kobayashi. Sample-based crowding method for multimodal optimization in continuous domain. In G. W. Greenwood, editor, *Proc. 2005 Congress on Evolutionary Computation (CEC'05)*. IEEE Press, 2005.
11. M. Appel and R. P. Russo. The connectivity of a graph on uniform points on 0, 1d. *Statistics & Probability Letters*, 60(4):351–357, 2002.
12. D. Ashlock, K. Bryden, and S. Corns. On taxonomy of evolutionary computation problems. In *Proceedings of the 2004 IEEE Congress on Evolutionary Computation*, pages 1713–1719. IEEE Press, 2004.
13. A. Auger and N. Hansen. Performance evaluation of an advanced local search evolutionary algorithm. In B. McKay et al., editors, *Proc. 2005 Congress on Evolutionary Computation (CEC'05)*. IEEE Press, 2005.
14. A. Auger and N. Hansen. A restart CMA evolution strategy with increasing population size. In B. McKay et al., editors, *Proc. 2005 Congress on Evolutionary Computation (CEC'05)*, pages 1769–1776. IEEE Press, 2005.
15. R. S. Barr, B. L. Golden, J. P. Kelly, M. G. Resende, and W. R. Stewart. Designing and reporting on computational experiments with heuristic methods. *Journal of Heuristics*, 1(1):9–32, 1995.
16. T. Bartz-Beielstein. *Experimental Research in Evolutionary Computation – The New Experimentalism*. Natural Computing Series. Springer, 2006.
17. T. Bartz-Beielstein. Neyman-Pearson theory of testing and Mayo's extensions in evolutionary computing. In A. Chalmers, D. R. Cox, C. Glymour, D. Mayo, and A. Spanos, editors, *First Symposium on Philosophy, History, and Methodology of E.R.R.O.R: Experimental Reasoning, Reliability, Objectivity & Rationality: Induction, Statistics, & Modeling. Virginia Tech, Blacksburg, Virginia, USA*, 2006.
18. T. Bartz-Beielstein, M. Chiarandini, L. Paquete, and M. Preuss. *Experimental Methods for the Analysis of Optimization Algorithms*. Springer, 2010.
19. T. Bartz-Beielstein, C. Lasarczyk, and M. Preuss. Sequential parameter optimization. In B. McKay et al., editors, *Proc. 2005 Congress on Evolutionary Computation (CEC'05)*, volume 1, pages 773–780. IEEE Press, 2005.

20. T. Bartz-Beielstein, C. Lasarczyk, and M. Preuss. The sequential parameter optimization tool-box. In T. Bartz-Beielstein, M. Chiarandini, L. Paquete, and M. Preuss, editors, *Experimental Methods for the Analysis of Optimization Algorithms*, pages 337–362. Springer, 2010.

21. T. Bartz-Beielstein, K. E. Parsopoulos, and M. N. Vrahatis. Design and analysis of optimization algorithms using computational statistics. *Applied Numerical Analysis & Computational Mathematics (ANACM)*, 1(2):413–433, 2004.

22. T. Bartz-Beielstein and M. Preuss. Considerations of budget allocation for sequential parameter optimization (SPO). In L. Paquete, M. Chiarandini, and D. Basso, editors, *Empirical Methods for the Analysis of Algorithms, Workshop EMAA 2006, Proceedings*, pages 35–40, 2006.

23. T. Bartz-Beielstein and M. Preuss. The future of experimental research. In T. Bartz-Beielstein, M. Chiarandini, L. Paquete, and M. Preuss, editors, *Experimental Methods for the Analysis of Optimization Algorithms*, pages 17–49. Springer, 2010.

24. T. Bartz-Beielstein and M. Preuss. Experimental analysis of optimization algorithms: Tuning and beyond. In Y. Borenstein and A. Moraglio, editors, *Theory and Principled Methods for the Design of Metaheuristics*. Springer, 2013.

25. T. Bartz-Beielstein, M. Preuss, and G. Rudolph. Investigation of one-go evolution strategy/quasi-Newton hybridizations. In F. Almeida et al., editors, *Hybrid Metaheuristics, Third International Workshop, HM 2006, Proceedings*, volume 4030 of *Lecture Notes in Computer Science*, pages 178–191. Springer, 2006.

26. D. Beasley, D. R. Bull, and R. R. Martin. A sequential niche technique for multimodal function optimization. *Evolutionary Computation*, 1(2):101–125, 1993.

27. R. Becker and G. Lago. A global optimization algorithm. In *Proceedings of the 8th Allerton Conference on Circuits and Systems Theory*, pages 3–12, 1970.

28. P. Berenbrink and T. Sauerwald. The weighted coupon collector's problem and applications. In H. Ngo, editor, *Computing and Combinatorics*, volume 5609 of *Lecture Notes in Computer Science*, pages 449–458. Springer, 2009.

29. P. Berkhin. Survey of clustering data mining techniques. Technical report, Accrue Software, San Jose, CA, 2002.

30. H.-G. Beyer et al. Evolutionary algorithms - terms and definitions. VDI/VDE guideline 3550, leaf 3, 2003. online version at http://ls11-www.cs.uni-dortmund.de/people/beyer/EA-glossary/def-engl-html.html.

31. H.-G. Beyer and H.-P. Schwefel. Evolution strategies: A comprehensive introduction. *Natural Computing*, 1(1):3–52, 2002.

32. K. S. Beyer, J. Goldstein, R. Ramakrishnan, and U. Shaft. When is "nearest neighbor" meaningful? In *ICDT '99: Proceeding of the 7th International Conference on Database Theory*, pages 217–235. Springer, 1999.

33. M. Birattari. *The Problem of Tuning Metaheuristics, as seen from a Machine Learning Perspective*. PhD thesis, Université Libre de Bruxelles, 2004.

34. M. Birattari, T. Stützle, L. Paquete, and K. Varrentrapp. A racing algorithm for configuring metaheuristics. In W. B. Langdon, E. Cantú-Paz, K. Mathias, R. Roy, G. Rudolph, et al., editors, *Proc. Genetic and Evolutionary Computation Conf. (GECCO 2002)*. Morgan Kaufmann, 2002.

35. M. Birattari, Z. Yuan, P. Balaprakash, and T. Stützle. F-race and iterated F-race: An overview. In T. Bartz-Beielstein, M. Chiarandini, L. Paquete, and M. Preuss, editors, *Experimental Methods for the Analysis of Optimization Algorithms*, pages 311–336. Springer, 2010.

36. S. Bird and X. Li. Adaptively choosing niching parameters in a PSO. In M. Cattolico, editor, *Genetic and Evolutionary Computation Conference, GECCO 2006, Proceedings*, pages 3–10. ACM, 2006.

37. C. Boender, A. Rinnooy Kan, G. Timmer, and L. Stougie. A stochastic method for global optimization. *Mathematical Programming*, 22(1):125–140, 1982.

38. K. D. Boese, A. B. Kahng, and S. Muddu. New adaptive multistart techniques for combinatorial global optimizations. *Operations Research Letters*, 16(2):101–113, 1994.

39. S. Boyd and L. Vandenberghe. *Convex Optimization*. Cambridge University Press, 2004.

40. H. J. Bremermann. Optimization through evolution and recombination. In M. Yovits, G. Jacobi, and G. Goldstein, editors, *Self-Organizing Systems*. Spartan Books, 1962.

41. R. Brits, A. P. Engelbrecht, and F. V. D. Bergh. A niching particle swarm optimizer. In *Proceedings of the 4th Asia-Pacific conference on simulated evolution and learning*, pages 692–696, 2002.
42. P. Calégari, G. Coray, A. Hertz, D. Kobler, and P. Kuonen. A taxonomy of evolutionary algorithms in combinatorial optimization. *Journal of Heuristics*, 5(2):145–158, 1999.
43. Catherine Cole McGeoch. *Experimental analysis of algorithms*. PhD thesis, Carnegie Mellon University, Pittsburgh, 1986.
44. G. Cattaneo and G. Italiano. Algorithm engineering. *ACM Comput. Surv.*, 31(3es):3, 1999.
45. CDF Collaboration, D0 Collaboration, and the Tevatron Electroweak Working Group. Combination of CDF and D0 results on the top-quark mass. Technical Report FERMILAB-TM-2323-E, Fermilab, IL, 2005.
46. M. Chimani, M. Kandyba, and M. Preuss. Hybrid numerical optimization for combinatorial network problems. In T. Bartz-Beielstein, M. J. B. Aguilera, C. Blum, B. Naujoks, A. Roli, G. Rudolph, and M. Sampels, editors, *Hybrid Metaheuristics, 4th International Workshop, HM 2007, Proceedings*, volume 4771 of *Lecture Notes in Computer Science*, pages 185–200. Springer, 2007.
47. M. Chimani and K. Klein. Algorithm engineering: Concepts and practice. In T. Bartz-Beielstein, M. Chiarandini, L. Paquete, and M. Preuss, editors, *Experimental Methods for the Analysis of Optimization Algorithms*, pages 131–158. Springer, 2010.
48. C. K. Chow and S. Y. Yuen. An evolutionary algorithm that makes decision based on the entire previous search history. *IEEE Trans. Evolutionary Computation*, 15(6):741–769, 2011.
49. V. Claus and A. Schwill. *Duden Informatik*. Duden-Verlag, 3rd edition, 2001.
50. C. A. Coello Coello. Theoretical and numerical constraint-handling techniques used with evolutionary algorithms: A survey of the state of the art. *Computer Methods in Applied Mechanics and Engineering*, 191(11–12):1245–1287, 2002.
51. C. A. Coello Coello and N. Cruz Cortés. Solving multiobjective optimization problems using an artificial immune system. *Genetic Programming and Evolvable Machines*, 6(2):163–190, 2005.
52. P. R. Cohen. A survey of the Eighth National Conference on Artificial Intelligence: Pulling together or pulling apart? *AI Magazine*, 12(1):16–41, 1991.
53. J. A. Coyne and H. A. Orr. *Speciation*. Sinauer Associates, 2004.
54. P. Dalgaard. *Introductory Statistics with R*. Springer, 2002.
55. S. Das, S. Maity, B.-Y. Qu, and P. N. Suganthan. Real-parameter evolutionary multimodal optimization - a survey of the state-of-the-art. *Swarm and Evolutionary Computation*, 1(2):71–88, 2011.
56. K. A. De Jong. *An analysis of the behavior of a class of genetic adaptive systems*. PhD thesis, University of Michigan, 1975.
57. K. A. De Jong. *Evolutionary Computation: A Unified Approach*. MIT Press, 2006.
58. K. A. De Jong. Parameter setting in EAs: a 30 year perspective. In F. G. Lobo, C. F. Lima, and Z. Michalewicz, editors, *Parameter Setting in Evolutionary Algorithms*. Springer, 2007.
59. K. A. De Jong, M. A. Potter, and W. M. Spears. Using problem generators to explore the effects of epistasis. In T. Bäck, editor, *ICGA*, pages 338–345. Morgan Kaufmann, 1997.
60. K. Deb and A. Saha. Multimodal optimization using a bi-objective evolutionary algorithm. *Evolutionary Computation*, 20(1):27–62, 2012.
61. A. Della Cioppa, C. De Stefano, and A. Marcelli. Where are the niches? Dynamic fitness sharing. *IEEE Trans. Evolutionary Computation*, 11(4):453–465, 2007.
62. C. Demetrescu, I. Finocchi, and G. Italiano. Algorithm engineering. *Bulletin of the EATCS*, (79):48–63, 2003.
63. P. J. Diggle. *Statistical Analysis of Spatial Point Patterns*. Arnold, 2nd edition, 2003.
64. S. Djebali et al. Landscape of transcription in human cells. *Nature*, 489(7414):101–108, 2012.
65. J. P. K. Doye. Physical perspectives on the global optimization of atomic clusters. In J. Pintér, editor, *Global Optimization — Scientific and Engineering Case Studies*, volume 85 of *Nonconvex Optimization and Its Applications*. Springer, 2006.

66. F. Drepper, R. Heckler, and H.-P. Schwefel. Ein integriertes System von Schätzverfahren, Simulations- und Optimierungstechniken zur Rechnergestützten Langfristplanung. In K.-H. Böhling and P. P. Spies, editors, *GI Jahrestagung*, volume 19 of *Informatik-Fachberichte*, pages 296–308. Springer, 1979. (in German).

67. S. Droste, T. Jansen, and I. Wegener. Upper and lower bounds for randomized search heuristics in black-box optimization. *Theory of Computing Systems*, 39(4):525–544, 2006.

68. S. Droste and D. Wiesmann. Metric based evolutionary algorithms. In *Proceedings of the European Conference on Genetic Programming*, pages 29–43. Springer, 2000.

69. D. Dumitrescu. Genetic chromodynamics. *Studia Universitatis Babes-Bolyai Cluj-Napoca, Ser. Informatica*, 45(1):39–50, 2000.

70. A. E. Eiben, R. Hinterding, and Z. Michalewicz. Parameter control in evolutionary algorithms. *IEEE Transactions on Evolutionary Computation*, 3(2):124–141, 1999.

71. A. E. Eiben and M. Jelasity. A critical note on experimental research methodology in EC. In *Proceedings of the 2002 Congress on Evolutionary Computation (CEC 2002)*, pages 582–587. IEEE Press, 2002.

72. A. E. Eiben and C. A. Schippers. On evolutionary exploration and exploitation. *Fundam. Inform.*, 35(1-4):35–50, 1998.

73. A. E. Eiben and M. Schoenauer. Evolutionary computing. *Information Processing Letters*, 82(1):1–6, 2002.

74. A. E. Eiben and J. E. Smith. *Introduction to Evolutionary Computing*. Springer, 2003.

75. M. T. M. Emmerich, A. H. Deutz, and J. W. Kruisselbrink. On quality indicators for black-box level set approximation. In *EVOLVE – A Bridge between Probability, Set Oriented Numerics and Evolutionary Computation*, volume 447 of *Studies in Computational Intelligence*, pages 157–185. Springer, 2013.

76. M. Epitropakis, V. Plagianakos, and M. Vrahatis. Finding multiple global optima exploiting differential evolution's niching capability. In *IEEE Symposium on Differential Evolution, 2011. SDE 2011. (IEEE Symposium Series on Computational Intelligence)*, pages 80–87. IEEE, 2011.

77. M. G. Epitropakis, X. Li, and E. K. Burke. A dynamic archive niching differential evolution algorithm for multimodal optimization. In *Proceedings of the 2013 IEEE Congress on Evolutionary Computation (CEC)*. IEEE, 2013.

78. B. S. Everitt, S. Landau, M. Leese, and D. Stahl. *Cluster Analysis*. Wiley Series in Probability and Statistics. Wiley, 5th edition, 2011.

79. D. G. Feitelson. Experimental Computer Science: The Need for a Cultural Change. http://www.cs.huji.ac.il/~feit/papers/exp05.pdf.

80. A. V. Fiacco and G. P. McCormick. *Nonlinear Programming: Sequential Unconstrained Minimization Techniques*. Research Analysis Corporation Research Series. Wiley, 1968.

81. R. A. Fisher. *The Design of Experiments*. Oliver and Boyd, 1935.

82. D. Fogel. Review of computational intelligence imitating life. *IEEE Transactions on Neural Networks*, 6(6):1562–1565, 1995.

83. D. B. Fogel. *Evolutionary Computation: The Fossil Record*. Wiley-IEEE Press, New York, 1998.

84. L. J. Fogel, A. J. Owens, and M. J. Walsh. Artificial intelligence through a simulation of evolution. In A. Callahan, M. Maxfield, and L. J. Fogel, editors, *Biophysics and Cybernetic Systems*. Spartan Books, 1965.

85. M. Gallagher and B. Yuan. A general-purpose tunable landscape generator. *IEEE Trans. Evolutionary Computation*, 10(5):590–603, 2006.

86. J. Gan and K. Warwick. A genetic algorithm with dynamic niche clustering for multimodal function optimisation. In *Artificial Neural Nets and Genetic Algorithms*, pages 248–255. Springer, 1999.

87. J. Gan and K. Warwick. Dynamic niche clustering: a fuzzy variable radius niching technique for multimodal optimisation in GAs. In *Proceedings of the 2001 Congress on Evolutionary Computation*, volume 1, pages 215–222, 2001.

88. J. Gan and K. Warwick. Modelling niches of arbitrary shape in genetic algorithms using niche linkage in the dynamic niche clustering framework. In *Proceedings of the 2002 Congress on Evolutionary Computation, CEC '02.*, volume 1, pages 43–48, 2002.

89. A. Giunta, S. Wojtkiewicz Jr., and M. Eldred. Overview of modern design of experiments methods for computational simulations. In *Proceedings of the 41st AIAA Aerospace Sciences Meeting and Exhibit.* American Institute of Aeronautics and Astronautics, 2003.

90. F. Glover and M. Laguna. *Tabu Search.* Kluwer Academic Publishers, 1997.

91. D. E. Goldberg and J. Richardson. Genetic algorithms with sharing for multimodal function optimization. In *Proceedings of the Second International Conference on Genetic Algorithms and Their Application*, pages 41–49. Lawrence Erlbaum Associates, 1987.

92. S. Gómez, N. del Castillo, L. Castellanos, and J. Solano. The parallel tunneling method. *Parallel Computing*, 29(4):523–533, 2003.

93. S. Gomez and A. Levy. The tunnelling method for solving the constrained global optimization problem with several non-connected feasible regions. In J. Hennart, editor, *Numerical Analysis*, volume 909 of *Lecture Notes in Mathematics*, pages 34–47. Springer, 1982.

94. P. Good. *Permutation, Parametric, and Bootstrap Tests of Hypotheses.* Springer, 3rd edition, 2005.

95. J. Grabmeier and A. Rudolph. Techniques of cluster algorithms in data mining. *Data Min. Knowl. Discov.*, 6(4):303–360, 2002.

96. J. Grefenstette. Optimization of control parameters for genetic algorithms. *IEEE Trans. Syst. Man Cybern.*, 16(1):122–128, 1986.

97. V. Hanagandi and M. Nikolaou. A hybrid approach to global optimization using a clustering algorithm in a genetic search framework. *Computers Chem. Engng.*, 22(12):1913–1925, 1998.

98. N. Hansen. The CMA evolution strategy: A tutorial. Accessed 13-08-12, http://www.lri.fr/ hansen/cmatutorial110628.pdf.

99. N. Hansen. Benchmarking a bi-population CMA-ES on the BBOB-2009 function testbed. In F. Rothlauf, editor, *Genetic and Evolutionary Computation Conference, GECCO 2009, Proceedings, Companion Material*, pages 2389–2396. ACM, 2009.

100. N. Hansen, A. Auger, S. Finck, and R. Ros. Real-parameter black-box optimization benchmarking 2009: Experimental setup. Research Report RR-6828, TAO–INRIA Saclay, Île de France, Paris, 2009.

101. N. Hansen, A. Auger, S. Finck, and R. Ros. Real-parameter black-box optimization benchmarking: Experimental setup, 2013. http://coco.lri.fr/downloads/download13.05/bbobdocexperiment.pdf, *accessed March 22, 2013.*

102. N. Hansen, S. Finck, R. Ros, and A. Auger. Real-parameter black-box optimization benchmarking 2009: Noiseless functions definitions. Research Report RR-6829, TAO–INRIA Saclay, Île de France, Paris, 2009.

103. N. Hansen and A. Ostermeier. Completely derandomized self-adaptation in evolution strategies. *Evolutionary Computation*, 9(2):159–195, 2001.

104. N. Hansen and R. Ros. Benchmarking a weighted negative covariance matrix update on the BBOB-2010 noiseless testbed. In *Proceedings of the 12th annual conference companion on Genetic and evolutionary computation*, GECCO '10, pages 1673–1680. ACM, 2010.

105. F. Hoffmeister and T. Bäck. Genetic algorithms and evolution strategies: Similarities and differences. Interner Bericht der Systems Analysis Research Group SYS–1/92, Universität Dortmund, Fachbereich Informatik, 1992.

106. J. H. Holland. Genetic algorithms and the optimal allocation of trials. *SIAM Journal of Computing*, 2(2):88–105, 1973.

107. R. Hooke and T. A. Jeeves. "Direct search" solution of numerical and statistical problems. *J. ACM*, 8(2):212–229, 1961.

108. J. Hooker. Testing heuristics: We have it all wrong. *Journal of Heuristics*, 1(1):33–42, 1996.

109. H. H. Hoos and T. Stützle. Evaluating Las Vegas algorithms: Pitfalls and remedies. In G. F. Cooper and S. Moral, editors, *UAI '98: Proc. of the 14th Conf. on Uncertainty in Artificial Intelligence*, pages 238–245. Morgan Kaufmann, 1998.

110. F. Hutter, H. H. Hoos, and K. Leyton-Brown. Sequential model-based optimization for general algorithm configuration. In C. A. Coello Coello, editor, *LION*, volume 6683 of *Lecture Notes in Computer Science*, pages 507–523. Springer, 2011.

111. F. Hutter, H. H. Hoos, K. Leyton-Brown, and T. Stützle. ParamILS: An automatic algorithm configuration framework. *Journal of Artificial Intelligence Research*, 36:267–306, 2009.

112. G. Jastrebski and D. Arnold. Improving evolution strategies through active covariance matrix adaptation. In *IEEE Congress on Evolutionary Computation, 2006. CEC 2006.*, pages 2814–2821, 2006.

113. M. Jelasity. UEGO, an abstract niching technique for global optimization. In A. E. Eiben, T. Bäck, M. Schoenauer, and H.-P. Schwefel, editors, *PPSN*, volume 1498 of *Lecture Notes in Computer Science*, pages 378–387. Springer, 1998.

114. D. S. Johnson. A theoretician's guide to the experimental analysis of algorithms. In M. H. Goldwasser, D. S. Johnson, and C. C. McGeoch, editors, *Data Structures, Near Neighbor Searches, and Methodology: Fifth and Sixth DIMACS Implementation Challenges*, pages 215–250. American Mathematical Society, 2002.

115. S. Kirkpatrick, C. D. Gelatt, and M. P. Vecchi. Optimization by simulated annealing. *Science, Number 4598, 13 May 1983*, 220, 4598:671–680, 1983.

116. J. E. Kobza, S. H. Jacobson, and D. E. Vaughan. A survey of the coupon collector's problem with random sample sizes. *Methodol Comput Appl Probab*, 9(4):573–584, 2007.

117. T. G. Kolda, R. M. Lewis, and V. J. Torczon. Optimization by direct search: New perspectives on some classical and modern methods. *SIAM Review*, 45(3):385–482, 2003.

118. J. R. Koza. *Genetic Programming: On the Programming of Computers by Means of Natural Selection*. MIT Press, 1992.

119. O. Kramer and H.-P. Schwefel. On three new approaches to handle constraints within evolution strategies. *Natural Computing*, 5(4):363–385, 2006.

120. T. Krink and M. Løvbjerg. The lifecycle model: Combining particle swarm optimisation, genetic algorithms and hillclimbers. In J. J. M. Guervós, P. Adamidis, H.-G. Beyer, J. L. Fernández-Villacañas, and H.-P. Schwefel, editors, *Parallel Problem Solving from Nature – PPSN VII, Proc. Seventh Intl. Conf.*, pages 621–630. Springer, 2002.

121. F. Kursawe. *Grundlegende empirische Untersuchungen der Parameter von Evolutions-strategien — Metastrategien*. Dissertation, Fachbereich Informatik, Universität Dortmund, 1999. in German.

122. J. C. Lagarias, J. A. Reeds, M. H. Wright, and P. E. Wright. Convergence properties of the Nelder–Mead simplex method in low dimensions. *SIAM Journal on Optimization*, 9(1):112–147, 1998.

123. W. B. Langdon and R. Poli. Evolving problems to learn about particle swarm and other optimisers. In B. McKay et al., editors, *Proc. 2005 Congress on Evolutionary Computation (CEC'05)*, volume 1, pages 81–88. IEEE Press, 2005.

124. M. Laumanns, G. Rudolph, and H.-P. Schwefel. A spatial predator-prey approach to multi-objective optimization: A preliminary study. In A. E. Eiben, M. Schoenauer, and H.-P. Schwefel, editors, *Parallel Problem Solving From Nature — PPSN V*, pages 241–249. Springer, 1998.

125. J.-P. Li, M. E. Balazs, G. T. Parks, and P. J. Clarkson. A species conserving genetic algorithm for multimodal function optimization. *Evolutionary Computation*, 10(3):207–234, 2002.

126. X. Li. Efficient differential evolution using speciation for multimodal function optimization. In H.-G. Beyer and U.-M. O'Reilly, editors, *Genetic and Evolutionary Computation Conference, GECCO 2005, Proceedings*, pages 873–880. ACM, 2005.

127. X. Li. A multimodal particle swarm optimizer based on fitness Euclidean-distance ratio. In H. Lipson, editor, *Genetic and Evolutionary Computation Conference, GECCO 2007, Proceedings*, pages 78–85. ACM, 2007.

128. X. Li. Niching without niching parameters: Particle swarm optimization using a ring topology. *IEEE Trans. Evolutionary Computation*, 14(1):150–169, 2010.

129. X. Li, A. Engelbrecht, and M. Epitropakis. Benchmark functions for CEC'13 special session and competition on niching methods for multimodal function optimization. Technical report, RMIT University, Evolutionary Computation and Machine Learning Group, Australia, 2013.

130. J. J. Liang, B.-Y. Qu, P. N. Suganthan, and A. G. Hernandez-Diaz. Problem definitions and evaluation criteria for the CEC 2013 special session and competition on real-parameter optimization. Technical Report 201212, Zhengzhou University and Nanyang Technological University, Singapore, 2013. http://www.ntu.edu.sg/home/EPNSugan.

131. I. Loshchilov, M. Schoenauer, and M. Sebag. Alternative restart strategies for CMA-ES. In C. A. Coello, V. Cutello, K. Deb, S. Forrest, G. Nicosia, and M. Pavone, editors, *Parallel Problem Solving from Nature - PPSN XII*, volume 7491 of *Lecture Notes in Computer Science*, pages 296–305. Springer, 2012.

132. I. Loshchilov, M. Schoenauer, and M. Sebag. Self-adaptive surrogate-assisted covariance matrix adaptation evolution strategy. In T. Soule and J. H. Moore, editors, *Genetic and Evolutionary Computation Conference, GECCO '12*, pages 321–328. ACM, 2012.

133. H. R. Lourenco, O. Martin, and T. Stützle. Iterated local search. In F. Glover and G. A. Kochenberger, editors, *Handbook of Metaheuristics*. Kluwer, 2002.

134. M. Lunacek, D. Whitley, and A. M. Sutton. The impact of global structure on search. In G. Rudolph, T. Jansen, S. M. Lucas, C. Poloni, and N. Beume, editors, *Parallel Problem Solving from Nature - PPSN X, 10th International Conference, Proceedings*, volume 5199 of *Lecture Notes in Computer Science*, pages 498–507. Springer, 2008.

135. R. I. Lung and D. Dumitrescu. A new evolutionary model for detecting multiple optima. In *Proceedings of the 9th annual conference on Genetic and evolutionary computation*, GECCO '07, pages 1296–1303. ACM, 2007.

136. S. W. Mahfoud. A comparison of parallel and sequential niching methods. In L. Eshelman, editor, *Proceedings of the Sixth International Conference on Genetic Algorithms*, pages 136–143. Morgan Kaufmann, 1995.

137. S. W. Mahfoud. *Niching methods for genetic algorithms*. PhD thesis, University of Illinois at Urbana-Champaign, 1995.

138. W. N. Martin, J. Lienig, and J. P. Cohoon. Island (migration) models: evolutionary algorithms based on punctuated equilibria. In T. Bäck, D. Fogel, and Z. Michalewicz, editors, *Handbook of Evolutionary Computation*, pages C6.3:1–C6.3:16. Inst. Phys. Publ., Oxford University Press, 1997.

139. R. L. Mason, R. F. Gunst, and J. L. Hess. *Statistical Design and Analysis of Experiments*. Wiley, 2nd edition, 2003.

140. D. G. Mayo. *Error and the Growth of Experimental Knowledge*. The University of Chicago Press, 1996.

141. E. Mayr. *Systematics and the Origin of Species*. Columbia University Press, 1942.

142. E. Mayr. Species, classification, and evolution. In R. Arai, M. Kato, and Y. Doi, editors, *Biodiversity and Evolution*. National Science Museum Foundation, Tokyo, 1995.

143. E. Mayr. *What Makes Biology Unique? — Considerations on the Autonomy of a Scientific Discipline*. Cambridge University Press, 2004.

144. G. P. McCabe and D. S. Moore. *Introduction to the Practice of Statistics*. W.H. Freeman, 2005.

145. C. McGeoch. *A Guide to Experimental Algorithmics*. A Guide to Experimental Algorithmics. Cambridge University Press, 2012.

146. M. D. McKay, R. J. Beckman, and W. J. Conover. A comparison of three methods for selecting values of input variables in the analysis of output from a computer code. *Technometrics*, 21(2):239–245, 1979.

147. O. Mersmann, B. Bischl, H. Trautmann, M. Preuss, C. Weihs, and G. Rudolph. Exploratory landscape analysis. In *Proceedings of the 13th annual conference on Genetic and evolutionary computation*, GECCO '11, pages 829–836. ACM, 2011.

148. O. Mersmann, M. Preuss, and H. Trautmann. Benchmarking evolutionary algorithms: Towards exploratory landscape analysis. In R. Schaefer, C. Cotta, J. Kolodziej, and G. Rudolph, editors, *Parallel Problem Solving from Nature, PPSN XI*, volume 6238 of *Lecture Notes in Computer Science*, pages 73–82. Springer, 2010.

149. N. Metropolis, A. W. Rosenbluth, M. N. Rosenbluth, A. H. Teller, and E. Teller. Equations of state calculations by fast computing machine. *J. Chem. Phys.*, 21:1087–1091, 1953.

150. Z. Michalewicz and M. Schoenauer. Evolutionary algorithms for constrained parameter optimization problems. *Evolutionary Computation*, 4(1):1–32, 1996.

151. B. L. Miller and M. J. Shaw. Genetic algorithms with dynamic niche sharing for multimodal function optimization. In *International Conference on Evolutionary Computation*, pages 786–791, 1996.

152. D. C. Montgomery. *Design and Analysis of Experiments*. Wiley, 5th edition, 2001.

153. B. M. Moret and H. D. Shapiro. Algorithms and experiments: The new (and old) methodology. *Journal of Universal Computer Science*, 7(5):434–446, 2001.

154. R. W. Morrison. Dispersion-based population initialization. In E. Cantú-Paz, J. A. Foster, K. Deb, L. D. Davis, R. Roy, U.-M. O'Reilly, H.-G. Beyer, et al., editors, *Proc. Genetic and Evolutionary Computation Conf. (GECCO 2003)*, volume 2723 of *Lecture Notes in Computer Science*, pages 1210–1221. Springer, 2003.

155. R. Motwani and P. Raghavan. *Randomized algorithms*. Cambridge University Press, 1995.

156. V. Nannen and A. E. Eiben. Relevance estimation and value calibration of evolutionary algorithm parameters. In M. M. Veloso, editor, *IJCAI 2007, Proceedings of the 20th International Joint Conference on Artificial Intelligence*, pages 975–980, 2007.

157. J. C. Nash. *Compact Numerical Methods for Computers: Linear Algebra and Function Minimisation*. IOP, 2nd edition, 1990.

158. J. Nelder and Mead. A simplex method for function minimization. *The Computer Journal*, (7):308–313, 1965.

159. A. S. Nemirovsky and D. B. Yudin. *Problem complexity and method efficiency in optimization*. Interscience series in discrete mathematics. Wiley, 1983. Translation of: Slozhnost' zadach i effectivnost' metodov optimizatsii, 1979.

160. F. J. Odling-Smee, K. N. Laland, and M. W. Feldman. *Niche Construction—The neglected process in evolution*. Princeton University Press, 2003.

161. F. Oppacher and M. Wineberg. The shifting balance genetic algorithm: Improving the GA in a dynamic environment. In W. Banzhaf, J. Daida, A. E. Eiben, M. H. Garzon, V. Honavar, M. Jakiela, and R. E. Smith, editors, *Proc. Genetic and Evolutionary Computation Conf. (GECCO 1999)*, volume 1, pages 504–510. Morgan Kaufmann, 1999.

162. D. Oshima, A. Miyamae, J. Sakuma, S. Kobayashi, and I. Ono. A new real-coded genetic algorithm using the adaptive selection network for detecting multiple optima. In *IEEE Congress on Evolutionary Computation, 2009. CEC '09*, pages 1912–1919. IEEE Press, 2009.

163. A. B. Owen. Latin supercube sampling for very high-dimensional simulations. *ACM Trans. Model. Comput. Simul.*, 8(1):71–102, 1998.

164. L. Pál, T. Csendes, M. C. Markót, and A. Neumaier. Black box optimization benchmarking of the global method. *Evolutionary Computation*, 20(4):609–639, 2012.

165. I. C. Parmee. The maintenance of search diversity for effective design space decomposition using cluster-oriented genetic algorithms (COGA) and multi-agent strategies (GAANT). In *Proceedings of 2nd International Conference on Adaptive Computing in Engineering Design and Control, ACEDC '96*, pages 128–138. University of Plymouth, 1996.

166. A. Pétrowski. A clearing procedure as a niching method for genetic algorithms. In T. Fukuda, T. Furuhashi, and D. B. Fogel, editors, *Proceedings of 1996 IEEE International Conference on Evolutionary Computation (ICEC '96)*, pages 798–803. IEEE Press, 1996.

167. A. Pétrowski. A classification tree for speciation. In P. J. Angeline and V. W. Porto, editors, *Proc. 1999 Congress on Evolutionary Computation (CEC'99)*, pages 204–211. IEEE Press, 1999.

168. M. Preuss. Niching prospects. In B. Filipic and J. Silc, editors, *Bioinspired Optimization Methods and their Applications (BIOMA 2006)*, pages 25–34. Jozef Stefan Institute, Ljubljana, 2006.

169. M. Preuss. Adaptability of algorithms for real-valued optimization. In M. Giacobini, A. Brabazon, S. Cagnoni, G. Caro, A. Ekárt, A. I. Esparcia-Alcázar, A. Farooq, A. Fink, and P. Machado, editors, *Applications of Evolutionary Computing*, volume 5484 of *Lecture Notes in Computer Science*, pages 665–674. Springer, 2009.

170. M. Preuss. Niching the CMA-ES via nearest-better clustering. In *Proceedings of the 12th annual conference companion on Genetic and evolutionary computation*, GECCO '10, pages 1711–1718. ACM, 2010.

171. M. Preuss. Improved topological niching for real-valued global optimization. In C. Di Chio et al., editors, *Applications of Evolutionary Computation*, volume 7248 of *Lecture Notes in Computer Science*, pages 386–395. Springer, 2012.

172. M. Preuss and T. Bartz-Beielstein. Sequential parameter optimization applied to self-adaptation for binary-coded evolutionary algorithms. In F. Lobo, C. Lima, and Z. Michalewicz, editors, *Parameter Setting in Evolutionary Algorithms*, Studies in Computational Intelligence, pages 91–120. Springer, 2007.

173. M. Preuss and C. Lasarczyk. On the importance of information speed in structured populations. In X. Yao, H.-P. Schwefel, et al., editors, *Parallel Problem Solving from Nature – PPSN VIII, Proc. Eighth Intl. Conf.*, pages 91–100. Springer, 2004.

174. M. Preuss, B. Naujoks, and G. Rudolph. Pareto set and EMOA behavior for simple multimodal multiobjective functions. In T. P. Runarsson et al., editors, *Parallel Problem Solving from Nature (PPSN IX)*, volume 4193 of *Lecture Notes in Computer Science*, pages 513–522. Springer, 2006.

175. M. Preuss, G. Rudolph, and F. Tumakaka. Solving multimodal problems via multiobjective technique with application to phase equilibrium detection. In D. Srinivasan and L. Wang, editors, *2007 IEEE Congress on Evolutionary Computation*, pages 2703–2710. IEEE Press, 2007.

176. M. Preuss, G. Rudolph, and S. Wessing. Tuning optimization algorithms for real-world problems by means of surrogate modeling. In M. Pelikan and J. Branke, editors, *Genetic and Evolutionary Computation Conference, GECCO 2010, Proceedings*, pages 401–408. ACM, 2010.

177. M. Preuss, L. Schönemann, and M. Emmerich. Counteracting genetic drift and disruptive recombination in $(\mu \overset{+}{,} \lambda)$-EA on multimodal fitness landscapes. In H.-G. Beyer, editor, *GECCO '05: Proceedings of the 2005 conference on Genetic and evolutionary computation*, pages 865–872. ACM Press, 2005.

178. M. Preuss, C. Stoean, and R. Stoean. Niching foundations: basin identification on fixed-property generated landscapes. In *Proceedings of the 13th annual conference on Genetic and evolutionary computation*, GECCO '11, pages 837–844. ACM, 2011.

179. M. Preuss, T. Wagner, and D. Ginsbourger. High-dimensional model-based optimization based on noisy evaluations of computer games. In Y. Hamadi and M. Schoenauer, editors, *Learning and Intelligent Optimization - 6th International Conference, LION 6*, volume 7219 of *Lecture Notes in Computer Science*, pages 145–159. Springer, 2012.

180. M. Preuss and S. Wessing. Measuring multimodal optimization solution sets with a view to multiobjective techniques. In *EVOLVE 2013*, Advances in Intelligent Systems and Computing (AISC). Springer, 2013.

181. K. Price. Differential evolution vs. the functions of the 2nd ICEO. In *Evolutionary Computation, 1997, IEEE International Conference on*, pages 153–157, 1997.

182. W. L. Price. A controlled random search procedure for global optimisation. *Computer Journal*, 20(4):367–370, 1977.

183. B. Qu, P. Suganthan, and S. Das. A distance-based locally informed particle swarm model for multimodal optimization. *Evolutionary Computation, IEEE Transactions on*, 17(3):387–402, 2013.

184. B.-Y. Qu, P. N. Suganthan, and J.-J. Liang. Differential evolution with neighborhood mutation for multimodal optimization. *IEEE Trans. Evolutionary Computation*, 16(5):601–614, 2012.

185. D. L. Rabosky, G. J. Slater, and M. E. Alfaro. Clade age and species richness are decoupled across the eukaryotic tree of life. *Plos Biology*, 10(8), 2012.

186. I. C. O. Ramos, M. C. Goldbarg, E. G. Goldbarg, and A. D. D. Neto. Logistic regression for parameter tuning on an evolutionary algorithm. In B. McKay et al., editors, *Proc. 2005 Congress on Evolutionary Computation (CEC'05)*, volume 2, pages 1061–1068. IEEE Press, 2005.

187. I. Rechenberg. *Evolutionsstrategie: Optimierung technischer Systeme nach Prinzipien der biologischen Evolution.* Frommann-Holzboog, 1973.

188. C. R. Reeves and T. Yamada. Genetic algorithms, path relinking, and the flowshop sequencing problem. *Evolutionary Computation*, 6(1):45–60, 1998.

189. A. Rinnooy Kan, C. Boender, and G. Timmer. A stochastic approach to global optimization. Technical Report WP1602-84, 1984.

190. J. Rönkkönen. *Continuous Multimodal Global Optimization with Differential Evolution-Based Methods.* PhD thesis, Lappeenranta University of Technology, 2009.

191. J. Rönkkönen, X. Li, V. Kyrki, and J. Lampinen. A framework for generating tunable test functions for multimodal optimization. *Soft Computing*, 15(9):1689–1706, 2011.

192. G. Rudolph. Self-adaptive mutations may lead to premature convergence. *IEEE Transactions on Evolutionary Computation*, 5(4):410–414, 2001.

193. T. P. Runarsson and X. Yao. Continuous selection and self-adaptive evolution strategies. In D. B. Fogel, M. A. El-Sharkawi, X. Yao, G. Greenwood, H. Iba, P. Marrow, and M. Shackleton, editors, *Proc. 2002 Congress on Evolutionary Computation (CEC'02)*, pages 279–284. IEEE Press, 2002.

194. J. Sacks, W. J. Welch, T. J. Mitchell, and H. P. Wynn. Design and analysis of computer experiments. *Statistical Science*, 4(4):409–435, 1989.

195. R. Salomon. Re-evaluating genetic algorithm performance under coordinate rotation of benchmark functions: A survey of some theoretical and practical aspects of genetic algorithms. *BioSystems*, 39:263–278, 1996.

196. M. E. Samples, M. J. Byom, and J. M. Daida. Parameter sweeps for exploring parameter spaces of genetic and evolutionary algorithms. In F. G. Lobo, C. F. Lima, and Z. Michalewicz, editors, *Parameter Setting in Evolutionary Algorithms*. Springer, 2007.

197. R. Schaefer, K. Adamska, and H. Telega. Genetic clustering in continuous landscape exploration. In *Engineering Applications of Artificial Intelligence (EAAI), Vol. 17*, pages 407–416. Elsevier, 2004.

198. L. Schönemann, M. Emmerich, and M. Preuss. On the extinction of evolutionary algorithm subpopulations on multimodal landscapes. *Informatica (Slowenia)*, 28(4):345–351, 2004.

199. M. Schonlau, W. J. Welch, and R. R. Jones. Global versus local search in constrained optimization of computer models. In *New Development and Applications in Experimental Design*, number 34 in IMS Lecture Notes, pages 11–25. Institute of Mathematical Statistics, 1998.

200. M. Schütz and J. Sprave. Application of parallel mixed-integer evolution strategies with mutation rate pooling. In *Proceedings of the fifth annual conference on Evolutionary Programming*, pages 345–354. MIT Press, 1996.

201. O. Schütze, X. Esquivel, A. Lara, and C. A. Coello Coello. Using the averaged Hausdorff distance as a performance measure in evolutionary multiobjective optimization. *IEEE Transactions on Evolutionary Computation*, 16(4):504–522, 2012.

202. H.-P. Schwefel. *Kybernetische Evolution als Strategie der experimentellen Forschung in der Strömungstechnik.* Diplomarbeit, Technische Universität Berlin, Hermann Föttinger–Institut für Strömungstechnik, März 1965.

203. H.-P. Schwefel. Experimentelle Optimierung einer Zweiphasendüse, part I. Technical Report No. 35 of the Project MHD–Staustrahlrohr 11.034/68, AEG Research Institute, Berlin, October 1968.

204. H.-P. Schwefel. *Evolutionsstrategie und numerische Optimierung.* Dr.-Ing. Dissertation, Technische Universität Berlin, Fachbereich Verfahrenstechnik, 1975.

205. H.-P. Schwefel. *Evolution and Optimum Seeking.* Sixth-Generation Computer Technology. Wiley, 1995.

206. O. M. Shir. Niching in evolution strategies. In H.-G. Beyer, editor, *GECCO '05: Proceedings of the 2005 conference on Genetic and evolutionary computation*, pages 865–872. ACM Press, 2005.

207. O. M. Shir. *Niching in Derandomized Evolution Strategies and its Applications in Quantum Control.* PhD thesis, Universiteit Leiden, 2008.

208. O. M. Shir. Niching in evolutionary algorithms. In G. Rozenberg, T. Bäck, and J. N. Kok, editors, *Handbook of Natural Computing*, pages 1035–1069. Springer, 2012.
209. O. M. Shir and T. Bäck. Niche radius adaptation in the CMA-ES niching algorithm. In T. P. Runarsson, H.-G. Beyer, E. K. Burke, J. J. M. Guervós, L. D. Whitley, and X. Yao, editors, *Parallel Problem Solving from Nature - PPSN IX, 9th International Conference*, volume 4193 of *Lecture Notes in Computer Science*, pages 142–151. Springer, 2006.
210. O. M. Shir, M. Emmerich, and T. Bäck. Adaptive niche radii and niche shapes approaches for niching with the CMA-ES. *Evolutionary Computation*, 18(1):97–126, 2010.
211. O. M. Shir, M. Preuss, B. Naujoks, and M. Emmerich. Enhancing decision space diversity in evolutionary multiobjective algorithms. In M. Ehrgott, C. Fonseca, X. Gandibleux, J.-K. Hao, and M. Sevaux, editors, *Evolutionary Multi-Criterion Optimization*, volume 5467 of *Lecture Notes in Computer Science*, pages 95–109. Springer, 2009.
212. G. Singh and K. Deb. Comparison of multi-modal optimization algorithms based on evolutionary algorithms. In M. Cattolico, editor, *Proc. Genetic and Evolutionary Computation Conf. (GECCO 2006)*, pages 1305–1312. ACM Press, 2006.
213. S.N. Lophaven and H.B. Nielsen and J. Søndergaard. DACE - a MATLAB Kriging toolbox. Technical Report IMM-TR-2002-12, IMM – Informatics and Mathematical Modelling, Technical University of Denmark, Lyngby, 2002.
214. A. R. Solow and S. Polasky. Measuring biological diversity. *Environmental and Ecological Statistics*, 1(2):95–103, 1994.
215. W. Spears. Simple subpopulation schemes. In A. V. Sebald and L. J. Fogel, editors, *Proc. Third Annual Conf. Evolutionary Programming (EP'94)*, pages 296–307. World Scientific, 1994.
216. J. Sprave. A unified model of non-panmictic population structures in evolutionary algorithms. In P. J. Angeline and V. W. Porto, editors, *Proc. 1999 Congress on Evolutionary Computation (CEC'99)*, volume 2, pages 1384–1391. IEEE Press, 1999.
217. W. Stadje. The collector's problem with group drawings. *Advances in Applied Probability*, 22(4):866–882, 1990.
218. C. Stoean, M. Preuss, R. Gorunescu, and D. Dumitrescu. Elitist generational genetic chromodynamics - a new radii-based evolutionary algorithm for multimodal optimization. In B. McKay et al., editors, *Proc. 2005 Congress on Evolutionary Computation (CEC'05)*, volume 2, pages 1839–1846. IEEE Press, 2005.
219. C. Stoean, M. Preuss, R. Stoean, and D. Dumitrescu. Disburdening the species conservation evolutionary algorithm of arguing with radii. In H. Lipson, editor, *Genetic and Evolutionary Computation Conference, GECCO 2007, Proceedings*, pages 1420–1427. ACM, 2007.
220. C. Stoean, M. Preuss, R. Stoean, and D. Dumitrescu. EA-powered basin number estimation by means of preservation and exploration. In G. Rudolph, T. Jansen, S. M. Lucas, C. Poloni, and N. Beume, editors, *Parallel Problem Solving from Nature - PPSN X, 10th International Conference*, volume 5199 of *Lecture Notes in Computer Science*, pages 569–578. Springer, 2008.
221. C. Stoean, M. Preuss, R. Stoean, and D. Dumitrescu. Multimodal optimization by means of a topological species conservation algorithm. *IEEE Transactions on Evolutionary Computation*, 14(6):842–864, 2010.
222. R. Storn and K. Price. Differential evolution – a simple and efficient heuristic for global optimization over continuous spaces. *J. of Global Optimization*, 11(4):341–359, Dec. 1997.
223. F. Streichert, G. Stein, H. Ulmer, and A. Zell. A clustering based niching EA for multimodal search spaces. In P. Liardet, P. Collet, C. Fonlupt, E. Lutton, and M. Schoenauer, editors, *Artificial Evolution, 6th International Conference, Evolution Artificielle, EA 2003*, volume 2936 of *Lecture Notes in Computer Science*, pages 293–304. Springer, 2004.
224. P. N. Suganthan, N. Hansen, J. J. Liang, K. Deb, Y.-P. Chen, A. Auger, and S. Tiwari. Problem definitions and evaluation criteria for the CEC 2005 special session on real-parameter optimization. Technical report, Nanyang Technological University, May 2005. http://www.ntu.edu.sg/home/EPNSugan.
225. R. Thomsen. Multimodal optimization using crowding-based differential evolution. In *IEEE Congress on Evolutionary Computation*, volume 2, pages 1382–1389, 2004.

226. G. Timmer. *Global Optimization: A Stochastic Approach*. PhD thesis, Erasmus Universiteit Rotterdam (Centrum voor Wiskunde en Informatica, Amsterdam), 1984.

227. A. Törn, M. Ali, and S. Viitanen. Stochastic global optimization: Problem classes and solution techniques. *Journal of Global Optimization*, 14(4):437–447, 1999.

228. A. A. Törn. Global optimization as a combination of global and local search. In *Proceedings of Computer Simulation Versus Analytical Solutions for Business and Economic Models*, Studies 17, pages 191–206. Gothenburg Business Administration, 1973.

229. A. A. Törn and A. Žilinskas. *Global Optimization*, volume 350 of *Lecture Notes in Computer Science*. Springer, 1989.

230. A. A. Törn and S. Viitanen. Topographical global optimization. In C. Floudas and P. Pardalos, editors, *Recent Advances in Global Optimization*, pages 384–398. Princeton University Press, 1992.

231. A. A. Törn and S. Viitanen. Topographical global optimization using pre-sampled points. *Journal of Global Optimization*, 5(3):267–276, 1994.

232. A. A. Törn and S. Viitanen. Iterative topographical global optimization. In C. Floudas and P. Pardalos, editors, *State of the Art in Global Optimization*, pages 353–363. Kluwer Academic Publishers, 1996.

233. A. Tovchigrechko and I. A. Vakser. How common is the funnel-like energy landscape in protein-protein interactions? *Protein Science*, 10(8):1572–1583, 2001.

234. T. Ulrich, J. Bader, and L. Thiele. Defining and optimizing indicator-based diversity measures in multiobjective search. In *Parallel Problem Solving from Nature, PPSN XI*, volume 6238 of *Lecture Notes in Computer Science*, pages 707–717. Springer, 2010.

235. T. Ulrich and L. Thiele. Maximizing population diversity in single-objective optimization. In N. Krasnogor and P. L. Lanzi, editors, *13th Annual Genetic and Evolutionary Computation Conference, GECCO 2011, Proceedings*, pages 641–648, 2011.

236. R. K. Ursem. Multinational evolutionary algorithms. In P. J. Angeline, editor, *Proceedings of the Congress of Evolutionary Computation (CEC-99)*, volume 3, pages 1633–1640. IEEE Press, 1999.

237. R. K. Ursem. *Models for Evolutionary Algorithms and Their Applications in System Identification and Control Optimization*. PhD thesis, University of Aarhus, 2003.

238. K. Veeramachaneni, T. Peram, C. Mohan, and L. Osadciw. Optimization using particle swarms with near neighbor interactions. *Genetic and Evolutionary Computation GECCO 2003*, pages 200–200, 2003.

239. J. Wakunda and A. Zell. Median-selection for parallel steady-state evolution strategies. In M. Schoenauer, K. Deb, G. Rudolph, X. Yao, E. Lutton, J. J. Merelo, and H.-P. Schwefel, editors, *Proc. Parallel Problem Solving from Nature – PPSN VI*, pages 405–414. Springer, 2000.

240. S. Wessing, M. Preuss, and G. Rudolph. When parameter tuning actually is parameter control. In N. Krasnogor and P. L. Lanzi, editors, *13th Annual Genetic and Evolutionary Computation Conference, GECCO 2011*, pages 821–828. ACM, 2011.

241. D. Whitley, S. B. Rana, J. Dzubera, and K. E. Mathias. Evaluating evolutionary algorithms. *Artificial Intelligence*, 85(1-2):245–276, 1996.

242. D. Whitley, J.-P. Watson, A. L. Howe, and L. Barbulescu. Testing, evaluation and performance of optimization and learning systems. Technical report, The GENITOR Research Group in Genetic Algorithms and Evolutionary Computation, Colorado State University, 2002.

243. Wikipedia. Category: Evolutionary algorithms. accessed June, 2006.

244. M. Wineberg. *Improving the Behavior of the Genetic Algorithm in a Dynamic Environment*. PhD thesis, Carleton University, Ottawa, Canada, 2000.

245. D. H. Wolpert and W. G. Macready. No free lunch theorems for optimization. *IEEE Transactions on Evolutionary Computation*, 1(1):67–82, April 1997.

246. M. H. Wright. Direct search methods: Once scorned, now respectable. In *Numerical Analysis 1995 (Proceedings of the 1995 Dundee Biennial Conference in Numerical Analysis)*, volume 344 of *Pitman Res. Notes Math. Ser.*, pages 191–208. CRC Press, 1995.

247. J. Yao, N. Kharma, and Y. Q. Zhu. On clustering in evolutionary computation. In *Congress on Evolutionary Computation, CEC 2006*, pages 1752–1759. IEEE Press, 2006.

248. X. Yin and N. Germay. A fast genetic algorithm with sharing using cluster analysis methods in multimodal function optimization. In *Proceedings of the International Conference on Artificial Neural Nets and Genetic Algorithms*, pages 450–457, 1993.

249. B. Yuan and M. Gallagher. Playing in continuous spaces: Some analysis and extension of population-based incremental learning. In R. Sarker et al., editors, *Proc. 2003 Congress on Evolutionary Computation (CEC'03)*, pages 443–450. IEEE Press, 2003.

250. B. Yuan and M. Gallagher. Combining meta-EAs and racing for difficult EA parameter tuning tasks. In F. G. Lobo, C. F. Lima, and Z. Michalewicz, editors, *Parameter Setting in Evolutionary Algorithms*. Springer, 2007.

251. Z. Zhai and X. Li. A dynamic archive based niching particle swarm optimizer using a small population size. In M. Reynolds, editor, *Thirty-Fourth Australasian Computer Science Conference, ACSC 2011*, volume 113 of *CRPIT*, pages 83–90. Australian Computer Society, 2011.

252. J. Zhang and A. Sanderson. JADE: Self-adaptive differential evolution with fast and reliable convergence performance. In *IEEE Congress on Evolutionary Computation, 2007. CEC 2007*, pages 2251–2258. IEEE Press, 2007.

Printed in the United States
By Bookmasters